Standard Methods for Thermal Comfort Assessment of Clothing

Textile Institute Professional Publications
Series Editor: The Textile Institute

Care and Maintenance of Textile Products Including Apparel and Protective Clothing
Rajkishore Nayak, Saminathan Ratnapandian

The Grammar of Pattern
Michael Hann

Standard Methods for Thermal Comfort Assessment of Clothing
Ivana Špelić, Alka Mihelić-Bogdanić, Anica Hursa Šajatović

For more information about this series, please visit: www.crcpress.com/Textile-Institute-Professional-Publications/book-series/TIPP

Standard Methods for Thermal Comfort Assessment of Clothing

Authored by
Ivana Špelić, Alka Mihelić-Bogdanić,
Anica Hursa Šajatović

CRC Press is an imprint of the
Taylor & Francis Group, an **informa** business

CRC Press
Taylor & Francis Group
6000 Broken Sound Parkway NW, Suite 300
Boca Raton, FL 33487-2742

© 2019 by Taylor & Francis Group, LLC
CRC Press is an imprint of Taylor & Francis Group, an Informa business

No claim to original U.S. Government works

Printed on acid-free paper

International Standard Book Number-13: 978-1-138-39147-5 (Hardback)
International Standard Book Number-13: 978-1-138-39098-0 (Paperback)

This book contains information obtained from authentic and highly regarded sources. Reasonable efforts have been made to publish reliable data and information, but the author and publisher cannot assume responsibility for the validity of all materials or the consequences of their use. The authors and publishers have attempted to trace the copyright holders of all material reproduced in this publication and apologize to copyright holders if permission to publish in this form has not been obtained. If any copyright material has not been acknowledged, please write and let us know so we may rectify in any future reprint.

Except as permitted under U.S. Copyright Law, no part of this book may be reprinted, reproduced, transmitted, or utilized in any form by any electronic, mechanical, or other means, now known or hereafter invented, including photocopying, microfilming, and recording, or in any information storage or retrieval system, without written permission from the publishers.

For permission to photocopy or use material electronically from this work, please access www.copyright.com (http://www.copyright.com/) or contact the Copyright Clearance Center, Inc. (CCC), 222 Rosewood Drive, Danvers, MA 01923, 978-750-8400. CCC is a not-for-profit organization that provides licenses and registration for a variety of users. For organizations that have been granted a photocopy license by the CCC, a separate system of payment has been arranged.

Trademark Notice: Product or corporate names may be trademarks or registered trademarks, and are used only for identification and explanation without intent to infringe.

Library of Congress Cataloging-in-Publication Data

Names: Špelić, Ivana, author. | Bogdanić, Alka Mihelić, author. | Šajatović, Anica Hursa, author.
Title: Standard methods for thermal comfort assessment of clothing / Ivana Špelić, Alka Mihelić Bogdanić, Anica Hursa Šajatović.
Description: First edition. | Boca Raton, FL : CRC Press/Taylor & Francis Group, 2018. | Includes bibliographical references and index.
Identifiers: LCCN 2018061308| ISBN 9781138391475 (hardback : alk. paper) | ISBN 9781138390980 (pbk. : alk. paper) | ISBN 9780429422997 (ebook)
Subjects: LCSH: Clothing and dress--Testing--Standards. | Textile fabrics--Testing--Standards. | Fashion design--Standards. | Clothing and dress--Physiological aspects. | Temperature--Physiological aspects. | Human comfort.
Classification: LCC TT499 .S64 2018 | DDC 646/.3--dc23
LC record available at https://lccn.loc.gov/2018061308

Visit the Taylor & Francis Web site at
http://www.taylorandfrancis.com

and the CRC Press Web site at
http://www.crcpress.com

*To our friends, families, colleagues and miracles ...
for you help us to keep learning, and without whom
none of this would have been possible.*

Contents

Preface xi
Textile Institute Professional Publications xiii
Author Bios xv

Chapter 1 Introduction 1

                References 4

Chapter 2 Behind the Scenes: Thermal Regulation in Humans 7

                2.1   Thermodynamical Analysis 8
                2.2   Physical Analysis 12
                      2.2.1   Heat Balance 13
                      2.2.2   Heat Production 15
                      2.2.3   Heat Loss 19
                2.3   Thermophysiological Analysis 23
                      2.3.1   The Human Thermal Homeostasis and
                              Thermostatic Neural Mechanism 23
                      2.3.2   The Anatomy of Thermoregulation 26
                              2.3.2.1   The Central Thermostatic Control and
                                        Hypothalamus 30
                              2.3.2.2   The Peripheral Reception and Skin 33
                References 38

Chapter 3 Modelling Heat Losses from the Human Body 43

                3.1   The Comfort in Humans 45
                3.2   Thermal Comfort: Environmental, Personal and Clothing
                      Properties 48
                      3.2.1   Clothing as a Second Skin: Preserving Thermal
                              Comfort 52
                      3.2.2   From Fibres to Clothing: Thermal Properties and
                              Applications 53
                              3.2.2.1   Heat Transmission through Textiles
                                        and Clothing 56
                            3.2.2.2   Moisture Transmission through
                                        Textiles and Clothing 62
                References 76

Chapter 4 The Importance of Globally Accepted Test Methods and
              Standards 83

                4.1   Testing the Thermal Properties of Textiles and Clothing 84

	4.1.1	Textile Thermal Comfort Testing..............................86
	4.1.2	Clothing Thermal Comfort Testing...........................91
	4.1.3	Subjective Judgements and Wear Trials: What Do We Have to Say?..97
4.2	Improving Comfort in Textiles and Clothing and Future Trends... 101	
References 112		

Chapter 5 Why Use Thermal Comfort Standards?.. 121

5.1 The Basic Principles of Standard Approval 123
5.2 The Types of Standards... 127
5.3 The Benefits Provided by Standards 128
References 130

Chapter 6 Who Creates Standards? 135

6.1 The National Organisations for Standardisation 135
    6.1.1 BSI Organisation 135
    6.1.2 DIN Organisation 136
    6.1.3 ASTM Organisation 136
    6.1.4 ASHRAE Organisation 137
6.2 The International Organization for Standardization (ISO) ... 138
References 141

Chapter 7 The Standardisation of Thermal Comfort... 145

7.1 The History of Standardisation 145
7.2 The Beginnings and First Standards of Thermal Comfort 146
7.3 Further Development and Necessary Improvements 154
References 163

Chapter 8 The Distribution of Standards on Thermal Comfort 173

8.1 The Basic Distributions of Standards in the Field of the Ergonomics of the Thermal Environment................... 174
8.2 Coding of Standards for Cataloguing Their Type 179
References 181

Chapter 9 Overview of the Most Significant Standards on Thermal Comfort 183

9.1 The Most Significant ISO and European Standards on Thermal Comfort.. 183
    9.1.1 The Standard Evaluating Physical Quantities........... 184
        9.1.1.1 Temperature Measurements – Requirements for Measuring Equipment.....186

			9.1.1.2	Humidity Measurements – Requirements for Measuring Equipment 189
			9.1.1.3	Pressure Measurements – Requirements for Measuring Equipment 190
			9.1.1.4	Air Velocity Measurements – Requirements for Measuring Equipment 191
		9.1.2	\multicolumn{2}{l	}{Standards Assessing the Thermal Comfort and Physiological Responses of Humans .... 192}
		9.1.3	\multicolumn{2}{l	}{The Standards Assessing Heat Stress .... 196}
		9.1.4	\multicolumn{2}{l	}{The Standards Assessing Cold Stress .... 202}
	9.2	\multicolumn{3}{l	}{The Most Significant ASTM Standards on Thermal Comfort .... 210}
	9.3	\multicolumn{3}{l	}{ASHRAE Handbook of Fundamentals .... 215}
	\multicolumn{4}{l	}{References .... 216}

Index 221
Standards Index 227

Preface

Sometimes we are unaware of how things can work out and how someone's faith in you can change the course of your everyday life. When we received the email with a proposal to write this book, there was no one happier than we were nor more afraid of the work that should be done in the future. You grow as a scientist reading about the great ideas written by people who had instilled you with love and admiration for both the knowledge they provided and the area of research. This book is merely a small portion of knowledge, donated by many of the researchers in the area of human thermal comfort, and our way of thanking them for their selfless work.

It was an honour and a privilege to write this book. The advances over the last few decades forced regulation improvements and international standards development in the area of thermal comfort. Since clothing is considered as one of the most important factors in modelling heat losses from the human body, thermal comfort assessment is a key element in guarding the human body while interacting with sometimes harsh environmental influences. Here is where the national and international standards provide help through the non-obligatory guidance and promotion of safe and reliable clothing thermal comfort assessment. For many decades, the study of human thermal comfort aimed to prevent health problems, cold-related injuries and thermal stress. Since comfort involves various contributing factors, both the psychological, physiological and physical, the study of thermal comfort assessment addresses many concerns. Since the beginnings of the study of the thermal comfort in humans, born in the early decades of the twentieth century, the multidisciplinary approach for defining problems and finding the optimal solutions is the only possible way. Through the selfless work of practitioners and scientists all over the world, we have found a way to cope with our environment. The development of standards and the unification of testing methods is of crucial significance for fostering thermal comfort research to serve us all in an objective, cost-reducing and health-protective manner.

Textile Institute Professional Publications

The aim of the *Textile Institute Professional Publications* is to provide support to textile professionals in their work and to help emerging professionals, such as final year or Master's students, by providing the information needed to gain a sound understanding of key and emerging topics relating to textile, clothing and footwear technology, textile chemistry, materials science and engineering. The books are written by experienced authors with expertise in the topic, and all texts are independently reviewed by textile professionals or textile academics.

The textile industry has a history of being both an innovator and an early adopter of a wide variety of technologies. There are textile businesses of some kind operating in all counties across the world. At any one time, there is an enormous breadth of sophistication in how such companies might function. In some places where the industry serves only its own local market, design, development and production may continue to be based on traditional techniques, but companies that aspire to operate globally find themselves in an intensely competitive environment, some driven by the need to appeal to followers of fast-moving fashion, others by demands for high performance and unprecedented levels of reliability. Textile professionals working within such organisations are subjected to a continued pressing need to introduce new materials and technologies, not only to improve production efficiency and reduce costs but also to enhance the attractiveness and performance of their existing products and to bring new products into being. As a consequence, textile academics and professionals find themselves having to continuously improve their understanding of a wide range of new materials and emerging technologies to keep pace with competitors.

The Textile Institute was formed in 1910 to provide professional support to textile practitioners and academics undertaking research and teaching in the field of textiles. The Institute quickly established itself as the professional body for textiles worldwide and now has individual and corporate members in over 80 countries. The Institute works to provide sources of reliable and up-to-date information to support textile professionals through its research journals, the *Journal of the Textile Institute*[1] and *Textile Progress*,[2] definitive descriptions of textiles and their components through its online publication *Textile Terms and Definitions*,[3] and contextual treatments of important topics within the field of textiles in the form of self-contained books such as the *Textile Institute Professional Publications*.

REFERENCES

1. http://www.tandfonline.com/action/journalInformation?show=aimsScope&journalCode=tjti20
2. http://www.tandfonline.com/action/journalInformation?show=aimsScope&journalCode=ttpr20
3. http://www.ttandd.org

Author Bios

Ivana Špelić, PhD, is a Postdoctoral Researcher at the University of Zagreb, Faculty of Textile Technology, Zagreb, Croatia. The areas of her special research interests are heat transfer in a body–clothing–environment system, applications of CAD systems for reverse engineering of human body modelling and thermal properties of clothing, protective clothing, energetics, technical thermodynamics and the application of alternative and renewable energy sources in textile and clothing industry.

Alka Mihelić-Bogdanić, PhD, is a Full Professor at the University of Zagreb, Faculty of Textile Technology, Zagreb, Croatia. The areas of her special research interests are heat recovery, energetics, technical thermodynamics, energy saving and management in industry as well as the application of alternative and renewable energy sources in the textile and clothing industry. She currently works as a member in the national research project involving thermophysiological properties of textile and footwear.

Anica Hursa Šajatović, PhD, is an Associate Professor at the University of Zagreb, Faculty of Textile Technology, Zagreb, Croatia. The areas of her special research interests are footwear production technology and protective footwear, protective clothing and clothing systems for protection from heat and fire and application of numerical modelling in garment production processes. She is a member of the technical committee of the Croatian Standards Institute in the area of protective clothing as well as leather and footwear.

1 Introduction

Vast knowledge of the principles of human physiology is required by practitioners for conducting most of the tests and measurements involving clothing. Since clothing is merely a part of the greater body–clothing–environment system, every single component of this system is connected and dependent on the other.

The heat and mass transfer pathways through textile and clothing structures are complex, and often a few simultaneous transport processes occur. Since clothing is a basic human need for survival, along with the food and shelter (*Textile Technology*, 2006), scientists and industry practitioners put a lot of effort into scientific research and technological improvements for producing textile structures and clothing for different end users and applications. Continuous improvements emerge on a daily basis for assisting with the thermal and moisture regulation of the human body through fibres, yarns and fabric construction engineering, and by developing special protective clothing for different environmental conditions and working environments.

New fabric and other textile materials are continuously designed in order to protect the human body from outside agents and to (*Heat & Mass Transfer in Textiles*, 2011):

1. Modify the water vapour and moisture loss rate of the insensible perspiration thus assisting the skin in conserving essential levels of body fluids or to cool the body.
2. Modify the heat loss rate of to keep the body in a cold environment at its critical internal temperature.
3. Keep the moisture penetration in the form of the rain and snow from the outside ambient to the skin and causing the body to become too cold.
4. Absorb solar hazardous electromagnetic waves and substances, such as ultraviolet radiation and toxic gases.
5. Block the penetration of harmful fluids such as blood-containing pathogens.

When it comes to clothing, its basic role is to modify the heat transfer between the human body and the environment. In this manner, the design process should always consider the intended usage and final purpose of the clothing. The other important factor to consider is the individual variations in heat exchange rate. The heat exchange rate varies greatly from person to person, for the various activities concerned and for the type of clothing worn (Gagge and Nishi, 1977). Finding common ground between the functional details needed, the production costs and the optimum design, which can provide the intended protection for the average consumer, involves the integration of various requirements in an all-around multidisciplinary approach.

In the past, testing of specialised protective clothing mainly concentrated on the material requirements and a few important clothing properties, although an important advantage of material test methods are that they are designed to produce accurate and reproducible results, both within and between laboratories, at minimal cost (Havenith and Heus, 2004).

Over the last few decades, clothing and textile design was driven by government and military needs for functional, protective and well-fitted garments, and performance-based garments for sport clothing companies. A new problem-solving design research area emerged focusing on both quantitative and qualitative methods in addition to practice (Bye, 2010).

Contemporary clothing design follows the crossover, multidisciplinary approach to clothing design involving the cooperation of two or more different areas for dealing with the more psychological needs of the complex survival environment. Growing clothing design requirements impose great diversification in the designs. The multidisciplinary design process incorporates human body engineering, medicine, chemical technology, nanotechnology, biotechnology, optics and many other disciplines. Besides the traditional people-oriented principle, the contemporary clothing design is fostered by new innovative material developments, the need for higher performance, development of intelligent clothing, the assistance of virtual technology and the need for sustainable and ecological development (Chunyan and Yue, 2014). Various effects need to be taken into account during the design development and testing of the protective clothing. These are the effects of the manufacturing process on the material's properties (stitching, seams, treatments), the effects of clothing design, sizing and fit, the effect of the interaction of the clothing with other components or gear and how the clothing performs in actual use (Havenith and Heus, 2004). The trend of developing multifunctional fashion garments, defined as clothing or clothing systems that allow adaptation to diverse social situations or weather conditions, or simply as having different characteristics for different body areas in order to have different functional features, resulted in new design requirements. One of the basic requirements of clothing design is the issue of comfort (Cunha and Broega, 2009).

In the light of thermophysiological comfort, clothing generally has to be designed to (Angelova, 2016; Vecchi et al., 2017):

1. Protect the human body from weather conditions.
2. Protect from cold and solar radiation.
3. Provide a protective barrier against heat flows (open fire and high temperatures).
4. Provide sufficient comfort in a physical and psychological sense, regardless of the type of activity.

The development of standards related to thermal comfort and the unification of testing methods are of crucial significance to the global market. This allows for clear communication of the product's performance, as well as allowing comparison between products and their materials from different producers. Information, requirements and quality control, which are described in international standards, provide applications for specific market segments and ensures that consumers' needs are fulfilled worldwide. Unlike regulations, which are general laws that cover an entire sector rather than a single product, standards are more complex. Regulations typically set out only a general framework while standards are voluntary but provide more detail on rules, guidelines or characteristics for a product or process. In the absence of an official international standard, one can also use standards from the national organisation for standardisation.

The current definitions of standards always emphasise their voluntary nature. The *Agreement on Technical Barriers to Trade* by the World Trade Organization defines a standard as a document approved by a recognised body *providing for common and repeated use rules, guidelines or characteristics for products or related processes and production methods compliance with which is not mandatory* (Agreement on Technical Barriers to Trade, 1995). Similar terminology is also visible in the ISO definition of standards. In *ISO/IEC Guide 2*, a standard is defined as a *document established by consensus* and approved by a recognised body, providing for common and repeated use rules, guidelines, or characteristics for activities or their results aimed at the achievement of the optimum degree of order in a given context. The guide defines standardisation as the activity of establishing – with regard to actual or potential problems – provisions for common and repeated use aimed at the achievement of the optimum degree of order in a given context (*ISO/IEC Guide 2*:2004).

The standards provide the establishment of unique and generally accepted guidelines, methods and regulations, which define the requirements for production processes, quality control and the product itself. They also provide a uniform approach to doing something through open access for the general good. The adoption of standards should support efficiency and overall cost reduction through competition while ensuring product quality, interoperability, safety and reliability. Standards represent a consensus between experts on the best way of doing something. They do not require an inventive step; rather, they document 'good practice' (Taylor and Kuyatt, 1994; Hatto, 2001).

There are a vast number of standards currently in usage around the globe, but only the ISO standards are considered as truly international standards. Since states and governments cannot be members, the ISO is best described as a centrally coordinated global network comprising hundreds of technical committees from all over the world and involving thousands of experts representing industries and other groups for developing and regularly maintaining technical standards. The ISO has grown into the world's largest and most widely recognised standards development organisation (O'Connor, 2015). The ISO's influence is exercised due to size, popularity and recognition. The World Trade Organization (WTO) placed the ISO with a status of the world's 'trade-legal' organisation (Morikawa and Morrison, 2004).

The ISO is formed of a global network of national standards bodies and serves as a global network of the world's leading standardisers. As a leading international organisation, the ISO is nowadays one of the most significant generators of contemporary analysing standards, although none of the international associations mentioned does the product testing itself. Most of the European national standards organisations are simultaneously members of the ISO such as the BSI (British Standards Institution), the DIN (German Institute for Standardization), the SNV (Swiss Association for Standardization), etc. There are also American national standards organisations, which market their products internationally, the ASTM and ANISI/ASHRAE. Thanks to the work performed by these organisations, nowadays there is around 19,500 ISO standards, 12,000 ASTM standards and a few regular ASHARE publications in usage (Špelić et al., 2016).

The advance in technology during and after the Industrial Revolution was accompanied by simultaneous progress in standards development, both in terms of scope and number. A great deal of standards similar to technical specifications emerged as a means of making mass production economically viable and globally applicable. The standardisation of production methods and products stimulated rapid business adaptation (*Standardization: Fundamentals, Impact, and Business Strategy,* 2010). However, standards should be continually reviewed in terms of new knowledge and future requirements to provide the best available methods for the future (Parsons, 2013).

There is still much work to do since there are almost no evident standards for clothing aimed at protecting against cold environments in comparison to standards for clothing aimed at protecting against heat and flames. Another evident flaw is the lack of any standards that address the requirements for everyday clothing, which doesn't fall in the scope of the protective clothing, but there is still great need to specify minimum requirements in order to produce optimum protection against cold.

This book is arranged into nine affiliated chapters. The first chapter is introductory and gives insight to the basic themes covered in the book. The second chapter provides detailed insights on human thermoregulation, from fundamental principles to different approaches in explaining the basic concepts of thermoregulation in humans. The focus is to explain the physiology of thermoregulation in humans that lies in the versatile description of the nervous system, temperature control mechanisms and energy transport modes. Chapter 3 covers the basic principles of a thermal comfort study, heat and moisture transmission through textiles and clothing, the function of clothing in preventing heat losses and factors influencing the ability to maintain a satisfactory thermal state. Chapters 4 to 9 discuss the details of thermal comfort standards: the globally accepted test methods for textile and clothing comfort testing; the basic principles and the significance of comfort standards; the methodology and reasons for creating standards; stakeholders, voting structure and standardising bodies; past, current and future development in this area; coding and naming the standards and last but not least everything you need to know on thermal comfort standards.

REFERENCES

Agreement on Technical Barriers to Trade. 1995. World Trade Organization, Final Act of the 1986–1994 Uruguay Round of trade negotiations.

Angelova, R. A. 2016. *Textiles and Human Thermophysiological Comfort in the Indoor Environment,* Boca Raton, FL: CRC Press, ISBN 978-1-4987-1539-3.

Bye, E. 2010. A direction for clothing and textile design research. *Clothing & Textiles Research Journal* 28(3): 205–217.

Chunyan, Q. and H. Yue. 2014. The research and development of the future fashion design. *American International Journal of Contemporary Research* 4(12): 126–130.

Cunha, J. and A. C. Broega. 2009. Designing multifunctional textile fashion products. In: *The Proceedings of the AUTEX 2009 World Textile Conference.* Izmir, Turkey: Ege University, 862–868.

Gagge, A. P. and Y. Nishi. 1977. Heat exchange between human skin surface and thermal environment, Chapter 5. In: *Handbook of Physiology, Supplement 26: Reactions to Environmental Agents,* ed. D. H. K. Lee, H. L. Falk and S. D. Murphy. Rockville, MD: American Physiological Society, 69–92, ISBN: 9780470650714.

Hatto. P. 2001. *Standards and Standardisation, A Practical Guide for Researchers.* European Commission, Directorate-General for Research & Innovation, Directorate G – Industrial technologies. https://ec.europa.eu/research/industrial_technologies/pdf/practical-standardisation-guide-for-researchers_en.pdf (accessed August 18, 2018).

Havenith, G. and R. Heus. 2004. A test battery related to ergonomics of protective clothing. *Applied Ergonomics* 35(1): 3–20.

Heat & Mass Transfer in Textiles. 2011. ed. A. K. Haghi, 2nd edition. Montreal, Canada: World Scientific and Engineering Academy and Society, WSEAS Press, ISBN: 978-1-61804-025-1.

ISO/IEC Guide 2:2004 Standardization and Related Activities – General Vocabulary. 2004. ISO – International Organisation for Standardization.

Morikawa, M. and J. Morrison. 2004. *Who Develops ISO Standards? A Survey of Participation in ISO's International Standards Development Processes.* Oakland, CA: Pacific Institute for Studies in Development, Environment, and Security. https://pacinst.org/reports/iso_participation/iso_participation_study.pdf (accessed August 18, 2018).

O'Connor. R. V. 2015. Developing software and systems engineering standards. In: *Proceedings of the 16th International Conference on Computer Systems and Technologies – CompSysTech'15*, ed. B. Rachev and A. Smrikarov. Dublin, Ireland.

Parsons, K. 2013. Occupational health impacts of climate change: Current and future ISO standards for the assessment of heat stress. *Industrial Health* 51: 86–100.

Špelić, I., D. Rogale and A. Mihelić-Bogdanić. 2016. An overview of measuring methods and international standards in the field of thermal environment, thermal characteristics of the clothing ensembles and the human subjects' assessment of the thermal comfort. *Tekstil* 65(3–4): 110–122.

Standardization: Fundamentals, Impact, and Business Strategy. 2010. APEC Sub Committee on Standards and Conformance Education Guideline 3 – Textbook for Higher Education. Singapore: Asia-Pacific Economic Cooperation (APEC). file:///C:/Users/spelici1/AppData/Local/Packages/Microsoft.MicrosoftEdge_8wekyb3d8bbwe/TempState/Downloads/210_cti_scsc_eduguide3_web.pdf (accessed on July 10, 2018).

Taylor, B. N. and C. E. Kuyatt. 1994. *NIST Technical Note 1297: Guidelines for Evaluating and Expressing the Uncertainty of NIST Measurement Results.* United States Department of Commerce Technology Administration: National Institute of Standards and Technology.

Textile Technology. 2006. ed. B. Wulfhorst, T. Gries and D. Veit. Munich, Germany: Carl Hanser Verlag, ISBN 1-56990-371-9.

Vecchi, R., R. Lamberts and C. M. Candido. 2017. The role of clothing in thermal comfort: How people dress in a temperature and humid climate in Brasil. *Ambiente Construído* 17(1): 69–81.

2 Behind the Scenes
Thermal Regulation in Humans

Humans are *homeotherms* (also called *endotherms*), so they have the ability to regulate their internal body temperature, unlike *poikilotherms* (also called *ectotherms*), which do not regulate their internal body temperature and depend on environmental temperature. In humans and homeothermic animals, there are two thermoregulation functions. First, the thermoregulation function counteracts external and internal temperature excitations, which are capable of temperature homeostasis and pose a danger for life. Second, the thermoregulation function serves to level out small but continuously arising internal and external temperature excitations (Ivanov, 2006).

The human body's internal body temperature is relatively stable over a broad range of ambient conditions. All of the energy contained in ingested food ultimately transforms into heat, and the rest to work performed on the environment or growth. The human thermoregulatory system creates an internal environment in which reaction rates are relatively high and avoids the pathologic consequences of wide deviations in body temperature. To maintain a relatively constant body temperature requires a balance between heat production and heat loss (Hulbert and Else, 2000; Boron and Boulpaep, 2012).

Due to the intrinsically high metabolic rates of their tissues, endotherms can maintain high and constant body temperatures (Hulbert and Else, 2000).

The body's thermoregulatory system consists of two intertwined components, a warm internal core and a cooler outer shell (mostly the skin). The body's core temperature is one of the most tightly regulated parameters of human physiology, but there are slight daily variations due to circadian rhythm, and monthly variations due to women's menstrual cycles (Kurz, 2008; Sessler, 2016). Humans regulate their internal body temperature, called the body's core temperature or temperature of the deep tissue, within a narrow range near 37°C, which usually remains constant within ±0.6°C, despite the variations in outside environmental temperature. The outer shell is not as strictly regulated within narrow limits because the environment greatly influences temperature alterations. In a warm environment, the shell may be less than 1 cm thick, but with humans conserving heat in a cold environment it may extend several centimetres below the skin (Witzmann, 2013). The skin temperature rises and falls, due to the environmental influences but is important for the ability to lose heat to the surroundings (Guyton and Hall, 2016).

The thermal interaction of the human body with the environment involves two simultaneous processes. One is heat transfer, which includes radiation, convection, conduction, evaporation and respiration, and the other is self-regulation in the form of vasoconstriction, vasodilation, shivering and sweating (Cheng et al., 2012). Aside from the subconscious mechanisms for body temperature control, the body uses behavioural control of temperature mechanisms (Guyton and Hall, 2016).

The thermoregulatory responses in cold environments serve to conserve heat within the body (cutaneous vasoconstriction) and to generate heat (shivering and non-shivering thermogenesis). Active regulation of body temperature is almost absent in thermoneutral environments, as heat production equals heat loss. In warm and hot environments, the human thermoregulatory mechanisms work to lose heat (cutaneous vasodilatation and sweating) (Charkoudian, 2016). There are two sources that contribute to heat accumulation inside the human body: internal heat (produced by the metabolism) and external heat (environmental heat), absorbed mainly due radiation and convection (Shapiro and Epstein, 1984).

Humans are more sensitive to danger from cold than from heat as presumed by the preponderance of cold spots over warm spots, and the shallower depth of cold spots relative to the skin surface (Arens and Zhang, 2006). Humans are able to thermal adapt to environments, due to prior thermal exposure affecting their thermal perception. Thermal adaptation significantly affects thermal comfort (Du et al., 2018).

2.1 THERMODYNAMICAL ANALYSIS

The word exergy was introduced by Slovenian scientist Zoran Rant (Kitanovski and Poredoš, 2016). Exergy is maximal technical amount of work, which can be obtained from the internal (intrinsic) energy of the substance. Exergy explains how much of the internal energy can be converted to the maximal work. The exergy is the measurement of the quantity of energy (Rant, 2001).

Energy analysis is a method of assessing the way energy is used in an operation involving the physical or chemical processing of materials and the transfer and/or conversion of energy, usually by performing an energy balance, based on the principle of the conservation of energy, and evaluating energy efficiencies. The exergy analysis method is based on both the 1st law of thermodynamics and the 2nd law of thermodynamics, indicating the locations of energy degradation in a process and quantifying the quality of heat in a waste stream. The exergy is the maximum theoretical useful work obtainable as the systems draw closer towards equilibrium, where heat transfer occurs only with the environment. Exergy depends on the properties of both a matter or energy flow and the environment, while the energy is independent of the environmental properties. This is a measure of the ability to do work of the great variety of streams (mass, heat, work) that flow through a system. Exergy equals zero when the system is in complete equilibrium with the environment. Exergy can be destroyed and generally is not conserved. The value of exergy can never be negative. Unlike energy, which cannot be destroyed nor produced (in accordance with the 1st law of thermodynamics), exergy can be neither destroyed nor produced in a reversible process but is always consumed in an irreversible process. Since energy is only a measure of quantity, exergy is a measure of both quantity and quality and appears in many forms (e.g. thermal exergy). It is measured on the basis of work or ability to produce work. Energy flows into and out of a system with mass flows, heat transfers and work interactions. Exergy can be transferred by exergy transfer associated with

Behind the Scenes

work, exergy transfer associated with heat transfer and exergy transfer associated with matter entering and exiting a control volume (Moran, 2000; Rant, 2001; Dincer and Rosen, 2007).

In order to explain human thermal comfort, the first condition that should be obtained is the existence of heat balance, based on the 1st law of thermodynamics. Exergy transfer and conversion accompany energy transfer and conversion in the human body. In other words, human energy transfer can be explained by the 1st and 2nd laws of thermodynamics (Prek, 2006). The human body can be described as a heat engine and is considered as an open system in a steady-state condition. The human thermal model could be thermodynamically expanded taking into account the 2nd law of thermodynamics (Prek et al., 2008). The human body produces heat and exchanges heat and mass at the boundary between the body surface and the environment (Prek and Butala, 2010).

Heat is defined as energy transported to or from a system due to a temperature difference between the system and its surroundings. Work is defined as a force moving through a distance. The only energy transport modes are heat, mass flow and work (Balmer, 2011).

The entropy is a measure of the system's disorder that is a property of the system's state. Entropy can only be produced (but not destroyed) within a system. The entropy of a system can be increased or decreased by entropy transport across the system boundary (Balmer, 2011). While the human body is exchanging heat with the environment, there are two boundaries. One boundary is between the core and the skin compartment and another between the skin compartment and the environment (Prek and Butala, 2010). The entropy of the isolated system continually increases, if the process occurs spontaneously, due to irreversible processes and reaches the maximum possible value when the system attains a state of thermodynamic equilibrium. The systems that exchange entropy with surroundings are not in an equilibrium state. An exchange of entropy is associated with the exchange of heat and matter (Dincer and Rosen, 2007). Entropy production equals exergy destruction in living systems. Entropy always increases in any non-equilibrium systems, following the 2nd law of thermodynamics. Exergy is consumed, and simultaneously entropy is produced as a consequence of heat and mass transfer or conversion. These processes are dependent on the human thermoregulatory system and on the state of the environment (Prek and Butala, 2010).

The exergy analysis method is a thermodynamic analysis, which explains the conservation of mass and energy in accordance with the 2nd law of thermodynamics. The 1st law of thermodynamics is the law of the conservation of energy. It states that energy cannot be created nor destroyed, although it can change form. The 2nd law of thermodynamics states that total entropy after a process is equal to or greater than that before (Dincer and Rosen, 2007). An open system is any system in which both mass and energy may cross the system boundary (Balmer, 2011).

The heat exchange between the human body and its environment follows the 2nd law of thermodynamics since it is proportional to the surface area and always occurs from a higher to a cooler temperature (Witzmann, 2013). Exergy (also called availability) is a measure of how useful the energy within a system can be. It is the amount of energy that is available to do useful work within a system. It is also a measure of the quality of the energy present within a system (Balmer, 2011). In order to establish the thermal comfort of the human body, three basic conditions must be fulfilled. The first is the existence of heat balance. The skin temperature and sweat rate must also be determined. Since the human body is an open system, it can exchange energy with the environment. The human body works to convert energy produced by metabolic processes into other forms, such as heat. The metabolic rate and the exergy of inhaled air are considered as dry heat inputs, while mass equal to water output (conservation of mass) is considered as wet exergy inputs. If the body is presented as a two-node model, with one node being the body's core and the second node being the skin layer, then the exchange of energy in the form of heat is directed passively through direct contact and dynamically through the thermoregulatory controlled peripheral blood flow (Prek et al., 2008).

The heat flow between the core and the skin of the human body can be written as (Prek et al., 2008):

$$\dot{Q}_{cr \to sk} = \left(K + \dot{m}_{bl} \times c_{bl}\right) \times \left(T_{cr} - \overline{T_{sk}}\right) \quad [W = J/s] \tag{2.1}$$

Where $\dot{Q}_{cr \to sk}$ is the rate of heat flow between the body core and the skin (W), K is the heat conduction from the core to the skin (W), $\dot{m}_{bl}$ is the mass flow rate of blood per unit of time (kg/s), c_{bl} is the specific heat of the blood (J/kgK), T_{cr} is the body core temperature (K) and $\overline{T_{sk}}$ is the temperature of the skin (K).

The exergy consumption is produced during the exergy transfer between the human body and the surrounding environment, if the human body's thermal transfer process is external irreversibility. The exergy consumption produced within the human body is assumed to be zero (Moran, 2000; Wua et al., 2013).

The human body exergy transfer is divided into five processes (Wua et al., 2013).

- The exergy generated by metabolism and work
- The exergy flow through respiration
- The exergy flow from evaporation
- The exergy flow from the skin to the surrounding environment
- The exergy stored in the core and skin compartments

The human body works to convert energy produced by the metabolism into the necessary forms taking into account other personal factors, such as clothing and activity for providing the desired thermal comfort level. Exergy input ($E_{x,in}$) consists of a dry heat input, which is equal to the metabolic rate plus the exergy of inhaled air, and of the wet exergy input, where its mass is equal to the water output (mass balance). The exergy output ($E_{x,out}$) is composed of dry exergy (determined by convective and radiative heat transfer from the skin, heat flow as the diffusion and evaporation of water from the skin surface, breathing) and wet exergy (diffusion and evaporation of water from the skin surface, air humidification while breathing). Since exergy measures the ability of energy to do work, it implies that exergy input ($E_{x,in}$) minus exergy consumption ($E_{x,cons}$) equals the exergy output ($E_{x,out}$), and using the exergy analysis the exergy consumption within the human body can be calculated (Prek, 2006; Prek et al., 2008).

Behind the Scenes

The general exergy balance of the human body can be written as (Prek et al., 2008):

$$E_{x,in} - E_{x,cons} = E_{x,stor} - E_{x,out} \quad [J] \tag{2.2}$$

Where $E_{x,in}$ is the input exergy (J), $E_{x,cons}$ is the exergy consumption (J), $E_{x,stor}$ is the stored exergy (J) and $E_{x,out}$ is the output exergy (J).

Since a steady-state condition is being assumed within the human body, there is no storage. If we presume that the heat storage is negligible, since the temperature and energy values can be assumed to be constant over time, the exergy balance for the human body can be written as (Prek, 2006; Prek et al., 2008):

$$c_{bl} \times \dot{m}_{bl} \times \left(T_{cr} - T_0 - T_0 \times \ln\frac{T_{cr}}{T_0}\right) + A_{Du} \times \varepsilon \times \sigma \times \left[T_{cr}^4 - T_0^4 - \frac{4}{3} \times \left(T_{cr}^4 - T_0^4\right)\right] + E_{x,cons}$$

$$= c_{bl} \times \dot{m}_{bl} \times \left(T_{sk} - T_0 - T_0 \times \ln\frac{T_{sk}}{T_0}\right) + A_{Du} \times h_c \times (T_{sk} - T_a) \times \frac{(T_{sk} - T_0)}{T_{sk}} \tag{2.3}$$

$$+ A_{Du} \times \varepsilon \times \sigma \times \left[T_{sk}^4 - T_0^4 - \frac{4}{3} \times \left(T_{sk}^4 - T_0^4\right)\right] \quad [J]$$

Where c_{bl} is the specific heat of the blood (J/kgK), $\dot{m}_{bl}$ is the mass flow rate of blood per unit of time (kg/s), T_0 is the reference temperature of the environment (K), T_{cr} is the body core temperature (K), T_{sk} is the temperature of the skin (K), T_a is the temperature of ambient air (K), h_c is the convective heat transfer coefficient (W/m²K), A_{Du} is the area of the human body calculated by the DuBois and Dubois formula (m²), ε is emissivity (–), σ is Stefan–Boltzmann constant (5.67 10^{-8} W/m²K⁴) and $E_{x,cons}$ is the exergy consumption (J).

DuBois body surface area (Dubois and Dubois, 1916; ASHRAE, 2017):

$$A_{Du} = 71.84 \times W_b^{0.425} \times H_b^{0.725} \quad [cm^2] \quad \text{or written as}$$

$$A_{Du} = 0.202 \times W_b^{0.425} \times H_b^{0.725} \quad [m^2] \tag{2.4}$$

Where A_{Du} is the area of the human body calculated by the DuBois and Dubois formula (m²), 0.007184 (apropos 0.202) is the constant, W_b is body weight (kg) and H_b is body height (m).

The left side of the equation represents exergy input ($E_{x,in}$), where the first term accounts for the thermal energy of the blood flow entering the skin compartment and the second term accounts for the radiation emitted by the surrounding surfaces. The right side of the equation represents the exergy output ($E_{x,out}$), which is explained as the thermal energy of the blood leaving the skin compartment, the convective heat transfer between the skin surface and the air, and the radiation emitted at the skin's surface. Under steady-state conditions, there is a connection between exergy consumption and expected levels of thermal comfort, showing that human exergy consumption is coupled to environmental conditions (Prek, 2005).

Most of the exergy analyses have been performed by assuming steady-state conditions for both the system and its environment (Choi et al., 2018). Optimal human performance coincides with minimum human body exergy consumption. Minimum human body exergy consumption is associated with a sensation close to thermal neutrality, inclining to slightly cool (Simone et al., 2011; Wua et al., 2013; Juusela and Shukuya, 2014). However, further studies indicate that minimal destroyed exergy does not correspond to thermal comfort conditions (Mady et al., 2014).

Exergy analysis is not yet applicable to adaptive thermal comfort models and should be further tested (Schweiker and Shukuya, 2012; Juusela and Shukuya, 2014). The minimum human body exergy consumption rate usually occurs when the mean radiant temperature is slightly higher than the mean air temperature (Juusela and Shukuya, 2014).

Since the exergy rates are strongly dependent on environmental conditions (Mady et al., 2012, 2014), the metabolism exergy is affected by the reference environment and core temperatures. The maximum exergy loss of the human body is proven to occur due to exhaled humid air, and thus most of the exergy is consumed by the human body in the summer season (Caliskan, 2013). The human body's exergy destruction under physical activity showed that 60% of the exergy is destroyed in the respiratory system (Albuquerque Neto et al., 2010). The energy transfer rates are larger than the exergy transfer rates when the subject is under physical activity (Mady et al., 2013).

The values of exergy flow rate due to evaporation for thermal comfort are smaller in the exergy method than in the conventional one during activity. Under physical activity, the exergy method for thermal comfort seems to be a reliable alternative to Fanger's PMV (predicted mean vote) and PPD (predicted percentage of dissatisfied) model. The rate of exergy destruction increases with activity intensity (Henriques et al., 2017).

Numerous studies continue to experiment with exergy analysis in order to evaluate the energy need for heating in indoor environments with regard to human thermal comfort models. They also work to extend the application of exergy analysis to other human thermal comfort models, except Fanger's (Dovjak et al., 2015; Buyak et al., 2017; Prek and Butala, 2017).

2.2 PHYSICAL ANALYSIS

Human homeostasis of internal temperature is maintained against wide variations of thermal flux from external or internal sources (Benzinger, 1964, 1965).

In order to explain the heat transfer between the human body and his environment, a simple physical analysis is performed. The biophysical basis for this interaction is when the thermal balance is observed in terms of the heat fluxes. The body temperature will remain steady as long as none of the physiological mechanisms that control human thermoregulation fail (Canadas, 2005).

The energy of the human body is exchanged with the environment. A part of it is exchanged as mechanical work, although most of the energy is exchanged in the form of the heat (Guyton and Hall, 2016). Heat is a by-product of all of the body tissues but is lost to the environment predominantly from the skin and from the respiratory tract. The heat transfers from heat producing sites to the rest of the body and afterwards from the body core to the outer shell. Heat is transported within the body by means of conduction (through the tissues) and convection (the flowing blood carries heat from warmer tissues to cooler tissues) (Witzmann, 2013).

The purpose of the human thermoregulatory system is to maintain a reasonably constant deep body temperature by maintaining a heat balance so that the heat lost to the environment is equal to the heat produced by the body. The human body possesses the most effective physiological mechanisms for maintaining a heat balance: sensible heat loss can be altered by a variation of the cutaneous blood flow and thus of the skin temperature, latent heat loss can be increased by sweat secretion, and internal heat production can be increased by shivering or muscle tension (Fanger, 1973a).

2.2.1 Heat Balance

The human thermoregulatory system that balances the gains and losses of energy for maintenance of constant temperature is capable of handling loads of four times a normal metabolic rate source (Benzinger, 1963). The transport of thermal energy (heat) in living tissue involves multiple mechanisms (conduction, convection, radiation, metabolism, evaporation and phase change). Thermal equilibrium is achieved after temperature stability had been achieved (Diller et al., 2000).

There are three basic heat exchange mechanisms in the human body – conduction, convection and radiation. Latent heat is exchanged through evaporation or in some cases through the condensation of water (Guyton and Hall, 2016).

The first comfort equation was derived by P. O. Fanger, based on the heat balance of the human body under certain boundary conditions, defining the factors that affect the human thermal comfort-environmental parameters (air temperature, mean radiant temperature, air velocity and humidity) and personal parameters (metabolic rate and clothing insulation) (Luo et al., 2016; d'Ambrosio Alfano et al., 2017). If the body is to remain at a relatively average constant temperature, then the heat outputs from the body must be equivalent to heat inputs to the body. There is a continuous heat exchange between the body and the environment. When the human body is in thermal equilibrium with negligible heat storage, the rate of heat generation equals the rate of heat loss, and the basic energy balance of the human body can be written as (Witzmann, 2013):

$$M - W = E + R + C + K + S \quad \left[W/m^2 \right] \tag{2.5}$$

Where (M–W) is the net heat production in the human body (W/m^2), M is the metabolic rate of the body (W/m^2), W is the rate of energy loss (W/m^2) as mechanical work (external work done by the muscles), E is the rate of heat loss by evaporation (W/m^2), R is the rate of heat loss by radiation (W/m^2), C is the rate of heat loss by convection (W/m^2), K is the rate of heat loss by conduction (W/m^2) and S is the rate of heat storage in the body (W/m^2), manifested as changes in tissue temperatures.

When the sum of energy production and energy gain from the environment does not equal energy loss, the extra heat is stored in the human body, or released from the body. The metabolic rate of the body (M) is always positive (Witzmann, 2013). The metabolic activity is the amount of energy needed for the body's basic functions when at rest – the need for energy and metabolic activity increases when working (Havenith, 1999). An adult person produces about 100 W of heat when at rest (Luo et al., 2018).

However, the quantities such as mechanical work (W), the rate of heat loss by evaporation (E), the rate of heat loss by radiation (R), the rate of heat loss by convection (C), the rate of heat loss by conduction (K) and the rate of heat storage in the body (S) can have a negative operator, since they represent energy exchange with the environment and storage. E, R, C, K and W are positive if they represent energy losses from the body and negative if they represent energy gains. When the rate of heat storage in the body (S) equals zero, the body is in heat balance, and body temperature neither rises nor falls. When the body is not in heat balance, its mean tissue temperature increases if S is positive and decreases if S is negative. Since the human body always tries to restore balance, the rate of heat storage in the body (S) will be positive or negative only until the body's responses to the temperature changes are insufficient to restore balance (Witzmann, 2013).

The energy balance equation for the human body can also be written, dividing most the variables into subcategories (*ASHRAE Handbook of Fundamentals*, 2017; Butera, 1998; Diller et al., 2000):

$$M - W = Q_{sk} + Q_{res} + S = (K + C + R + E_{sk}) + (C_{res} + E_{res}) + (S_{sk} + S_{cr}) \quad \left[W/m^2\right] \quad (2.6)$$

Where M is the rate of metabolic heat production (W/m^2), W is the rate of mechanical work accomplished, also called the external work (W/m^2), Q_{sk} is the total rate of heat loss from skin (W/m^2). The heat loss from the surface of the skin is the combination of the surface loss by convection, conduction, radiation and evaporation. K is the heat loss by conduction (W/m^2). C+R is the sensible heat loss from skin, which is a combination of the heat loss by convection and radiation (W/m^2) and E_{sk} is the total rate of evaporative heat loss from skin (W/m^2). Q_{res} is the total rate of heat loss through respiration (W/m^2). The respiratory heat loss is the combination of convection and evaporation. C_{res} is the rate of convective heat loss from respiration (W/m^2), and E_{res} is the rate of evaporative heat loss from respiration (W/m^2). S is the total rate of energy stored in the human body (W/m^2), S_{cr} is the rate of heat storage in core compartment (W/m^2), and S_{sk} is the rate of heat storage in the skin compartment (W/m^2).

The measured environmental variables affecting the heat-balance equation are ambient temperature, mean radiant temperature or the effective radiant field, the ambient vapour pressure or dew-point temperature, an angle factor that describes body posture and orientation to thermal radiation present, air movement as it affects the convective heat transfer coefficient, clothing insulation, barometric pressure and length of exposure. The physiological variables affecting the heat-balance equation are skin temperature, skin wettedness, the rate of change of body heat storage or of mean body temperature, the metabolic energy level along with the metabolic heat flowing through the skin surface, and the external work accomplished (Gagge and Nishi, 1977; Guyton and Hall, 2016).

When the muscles burn these nutrients for mechanical activity, the energy is liberated as external work outside the body, and the greater portion is released as heat. The ratio between this external work and the energy consumed is called the efficiency with which the body performs the work (Havenith, 1999). In the heat-balance equation, the conduction (K) is restricted to heat flow between the body and other solid objects, and it usually represents only a small part of the total heat exchange with the environment (Witzmann, 2013).

This energy balance equation between the human body and the uniform environment was developed on the basis of Povl Ole Fanger's work and his thermal comfort model (Fanger, 1970, 1973b). Fanger's work was accepted in the most frequently cited thermal comfort standards, ASHRAE 55-2017 (ANSI/ASHRAE Standard 55, 2017) and ISO 7730 (ISO 7730, 2005). P. O. Fanger described the energy balance as being the basis of general thermal comfort research (Orosa, 2009).

Heat moves from the skin to the environment by radiation, conduction, convection and evaporation. When the heat gain exceeds the heat loss, the body core temperature rises. In contrast, when the heat loss exceeds the heat gain, the body core temperature will start to decrease (Boron and Boulpaep, 2012).

If the heat production by metabolic rate is higher than the sum of all heat losses, the total rate of energy stored in the human body (S) will be positive, which means body heat content increases and body temperature rises. If the store is negative, more heat is lost than produced and the body cools (Havenith, 1999).

A person will come into a state of heat stress when their environment (air temperature, radiant temperature, humidity and air velocity), clothing and activity interact to produce a tendency for body temperature to rise. The heat stress will cause a transient or persistent imbalance between heat gain and heat loss, resulting in body heat storage (Parsons, 1998; Cramer and Jay, 2016). Cold stress will occur when a thermal load on the body produces a greater than normal heat loss, and compensatory thermoregulatory actions are required to maintain the body thermally neutral. When thermoregulation is impaired, and the core temperature starts to decline, the individual suffers cold stress (Holmér et al., 1998).

2.2.2 Heat Production

The human body uses food as a source of energy. After food intake, the body transforms the food by combustion and synthesises the adenosine triphosphate (ATP), a molecule considered as the energy currency in cell metabolism. The ATP is a nucleotide composed of nitrogenous base adenine, pentose sugar ribose and three phosphate radicals. Up to 95% of ATP is formed inside the mitochondria, filled with a matrix that contains large quantities of dissolved enzymes, by a chemiosmotic mechanism. Simultaneous cooperation between dissolved and oxidative enzymes cause oxidation of the food nutrients and the release of energy. This energy is used to synthesise the ATP. The adenosine triphosphate (ATP) has a large quantity of free energy – about 50.2 kJ/mol (12 kcal/mol) is liberated during ATP breakdown. This energy is needed for the active transport of molecules across cell membranes and synthesis of chemical compounds throughout the cell, the contraction of muscles and performance of the mechanical work, the glandular secretion and the synthetic reactions that create hormones, the maintenance of the membrane potentials by the nerve and muscle fibres to conduct the impulses, cell division and growth and the absorption of foods from the gastrointestinal tract (Witzmann, 2013; Guyton and Hall, 2016).

The principal by-product of human metabolism is heat. The rate of heat production in the human organism is called *the metabolic rate of the body*. The metabolism of the body is a common denominator for all of the chemical reactions in all the cells of the body, whereas the metabolic rate is normally expressed in terms of the rate of heat liberation during chemical reactions. When the human body isn't performing the work outside the body, and the external expenditure of energy is not taking place, all energy released by the metabolic processes eventually becomes body heat. Only a small portion of the energy from food transfers into adenosine triphosphate (ATP) production used for body function. During the phosphorylation of adenosine diphosphate (ADP) to form adenosine triphosphate (ATP) only about 43% is captured in the ATP, and the remaining 57% of the energy is lost in the form of heat. Only 27% of the produced energy from food is used by the functional systems of the cells, but even when this portion of the energy reaches the functional systems of the cells the rest of this energy eventually becomes heat (Guyton and Hall, 2016).

The conversion of metabolic energy into heat within the body may occur almost immediately as with the energy used for active transport or heat produced as a by-product of muscular activity. The energy used in forming glycogen or protein is released as heat with a delay (Witzmann, 2013).

Since the metabolism of the body simply means all the chemical reactions in all the cells of the body, the metabolic rate is the rate of heat liberation during chemical reactions. The unit of a quantity of energy, heat or metabolic energy in the human body is usually a kilojoule (kJ), kilocalorie (kcal) or watt (W). Direct calorimetry measures the total quantity of heat liberated from the body (equal to the metabolic rate). But since the direct calorimetry is difficult to perform, most of the time the indirect calorimetry is used by obtaining the energy equivalent of oxygen, called the rate of oxygen utilisation. This method is based on measuring a person's rate of oxygen consumption. More than 95% of the energy expended in the body is derived from reactions of oxygen with the different foods. When carbohydrates are metabolised per litre of oxygen consumed, around 21.14 kJ (5.05 kcal) is released, when metabolising fat, 19.67 kJ (4.70 kcal) is liberated, and 20.20 kJ (4.60 kcal) is released with protein combustion (Witzmann, 2013; Guyton and Hall, 2016). The quantity of energy liberated per litre of oxygen used in the body averages at about 4.825 kcal/L (20.2 kJ/L), which is called the energy equivalent of oxygen. The ratio of carbon dioxide produced to oxygen consumed in the tissues is called the respiratory quotient (RQ). The RQ is 1.0 for the oxidation of carbohydrates, 0.71 for the oxidation of fat, and 0.80 for the oxidation of protein. For a normal diet involving carbohydrates, fats and proteins, the average value of respiratory quotient is 0.825. The respiratory exchange ratio (R) is calculated when the rate of carbon dioxide output by the lungs is divided by the uptake of oxygen during the same period. During a period of 1 hour, the respiratory exchange ratio (R) is almost equal to the RQ under normal resting conditions (a steady state) (Witzmann, 2013; Guyton and Hall, 2016).

The metabolic rate measured under a set of standard conditions is called the basal metabolic rate (BMR) and presents the minimum energy expenditure for the body. It accounts for about 50 to 70% of the daily energy expenditure in most sedentary persons. The BMR is usually determined by measuring the rate of oxygen utilisation over a given period under the comfortable ambient temperature of 293.15 to 299.65 K (20 to 26.5°C) without physical activity during the test. Other requests include that the person fasted for 12 hours, slept well and rested without strenuous activity for at least 1 hour before the test. The metabolic rate during sleep is somewhat lower than the BMR. An average resting man weighing 70 kg uses the daily energy requirement of about 8374 kJ (2000 kcal). The metabolic rate of the body at rest varies according to body surface area, age, sex, hormones, special disorders such as thyroid gland disorders and digestion. In a resting and fasting young adult man with a body surface area of 1.8 m^2, the metabolic rate is about 45 W/m^2 (or 81 W or 70 kcal/h) and corresponding to an O_2 consumption of about 240 mL/min. The BMR value is therefore corrected for differences in body size since the skeletal muscles account for 20 to 30% of the BMR. Even though the body core comprises only about 36% of the overall body mass, it accounts for around 70% of the energy production at rest and the heat needed to maintain heat balance at comfortable environmental temperatures. A nonpregnant woman's metabolic rate is 5 to 10% lower than that of a man of the same age and surface area probably because a higher proportion of the female body is composed of fat, a tissue with a low metabolism. The increase in the metabolic rate after food digestion is called *the thermogenic effect of food*, because the chemical reactions associated with digestion, absorption and storage of food in the body require energy and generate heat. After food intake, metabolic rate increases 4 to 30% above normal at rest. After carbohydrate or fat digestion, the metabolic rate increases about 4%. After protein digestion, the metabolic rate begins rising within an hour, reaching a maximum of about 30% and lasting for 3 to 12 hours. The thermogenic effect of food accounts for about 8% of the total daily energy expenditure in many persons. Beside the thermogenic effect of food, there are more important ways of producing the body heat – performing physical work, shivering and *non-shivering thermogenesis* (Witzmann, 2013; Guyton and Hall, 2016).

The overall metabolic rate of the body is the combination of the following factors (Guyton and Hall, 2016):

- The basal metabolic rate.
- The increase in the metabolic rate caused by muscle activity (including shivering).
- The extra metabolism caused by hormones (the maximal amounts of thyroxine secreted by the thyroid gland, causes the metabolic rate to rise 50 to 100% above normal; the loss of thyroid secretion decreases the metabolic rate from 40 to 60% of the normal rate; the male sex hormone testosterone can increase the metabolic rate by about 10 to 15%; the growth hormone can increase the metabolic rate by stimulating cellular metabolism and by increasing skeletal muscle mass).

- The extra metabolism caused by the sympathetic stimulation and the effect of epinephrine and norepinephrine (the non-shivering thermogenesis).
- The extra metabolism caused by increased chemical activity in the cells and fever (fever increases the chemical reactions of the body by an average of 120% for every 10-degree rise in temperature).
- The extra metabolism needed for the thermogenic effect of food.

The mechanisms of increasing body temperature when it is too cold can be classified as (Guyton and Hall, 2016; Morrison, 2016):

- *Cutaneous vasoconstriction of skin blood vessels*: The vasoconstriction of the skin blood vessels is triggered by the stimulation of sympathetic centres (in the posterior hypothalamus that causes vasoconstriction).
- *Piloerection of the hair follicles*: Piloerection is triggered by sympathetic stimulation causing the arrector pili muscles attached to the hair follicles to contract and bring the hairs to an upright stance. The hairs entrap a thick layer of insulator air next to the skin, so a transfer of heat to the surroundings is greatly depressed, but this mechanism is less important in humans.
- *Increase in thermogenesis (heat production)*: Heat production by the metabolic systems is increased by promoting shivering, sympathetic excitation of heat production and thyroxine secretion. The principal sources of metabolic heat production beyond those contributing to basal metabolic rate are shivering in skeletal muscle and brown adipose tissue thermogenesis.

Shivering produces heat by increasing muscle activity in response to cold stress. The primary motor centre for shivering is located in the dorsomedial portion of the posterior hypothalamus. This centre activates when the core temperature falls even a fraction of a degree below a critical temperature level (set-point temperature of 310.15 K, which corresponds to 37°C) and transmits signals that cause shivering through the increase the tone of the skeletal muscles. The signals travel through bilateral tracts down the brain stem, into the lateral columns of the spinal cord and finally to the anterior motor neurons. During maximum shivering, body heat production can rise from four to five times its normal level (Guyton and Hall, 2016). There are two mechanisms of the chemical thermogenesis involved in heat production inside the human body. The first is the consequence of sympathetic chemical excitation causing chemical thermogenesis (*non-shivering thermogenesis*; an increase in the rate of cellular metabolism), due to an increase in either sympathetic stimulation or circulating norepinephrine and epinephrine in the blood. The excess food is oxidised to release heat but without adenosine triphosphate (ATP) formation. Since adult humans have almost no brown fat, the chemical thermogenesis can increase the rate of heat production for only 10 to 15% (Guyton and Hall, 2016). The second mechanism of the chemical thermogenesis involves thyroxine secretion. When the anterior hypothalamic-preoptic area is cooled, the hypothalamus starts releasing the neurosecretory hormone thyrotropin-releasing hormone. This hormone is carried by way of the hypothalamic portal veins to the anterior pituitary gland, where it stimulates secretion of a thyroid-stimulating hormone, which stimulates an increased output of thyroxine by the thyroid gland. Thyroxine activates uncoupling protein and increases the rate of cellular metabolism throughout the body. But this mechanism is slow and involves slow exposure to cold (Guyton and Hall, 2016).

The skeletal muscles are the principal source of metabolic heat during mild exercise, and even more so during intense exercise, when they may account for up to 90% of the heat production. During moderate exercise, a healthy sedentary young man requires a metabolic rate of 600 W. During intense exercise by a trained athlete, a metabolic rate of 1400 W or more is needed. The blood perfusing the muscles during activity is warmed because they are almost 1 degree warmer than the core of the body. Further circulation of this blood warms the rest of the body and raises the core temperature. Most of the energy in muscles is converted into heat rather than mechanical work. During muscle contraction, some of the energy in the ATP is again converted into heat rather than mechanical work. During muscle contraction, energy is required to perform work and in order to gain this energy large amounts of ATP are disintegrated to ADP (called the *Fenn effect*). No more than 25% of the input energy to the muscles (the chemical energy in nutrients) can be converted into mechanical work outside the body, and the other 75% or more is converted into heat within the body. This is called the efficiency of muscle contraction (Witzmann, 2013; Guyton and Hall, 2016).

2.2.3 Heat Loss

Several pathways are available for heat loss from the body, which are conduction, radiation, convective and evaporative heat loss from the skin and by respiration. A minor role is taken by conduction, whereas the most important avenue of the heat loss is convection since the surrounding air is usually cooler than the skin, and the heat is transferred from the skin to the surrounding air. Heat transfer by radiation will occur when there is a difference between the body's surface temperature and the temperature of the surfaces in the environment. There is also the heat loss by evaporation since the body has the ability to sweat and moisture appearing on the skin can evaporate, dissipating large amounts of heat from the body. The convective and evaporative heat loss occurs during respiration, as lungs warm and moisturise the inhaled air followed by heat loss with the expired air, which can be up to 10% of the total heat production (Fiala et al., 1999; Havenith, 1999).

Convection is the transfer of heat by the movement of a fluid, whereas in the human body the fluid is usually air or water in the environment; in the case of heat transfer inside the body, the fluid is blood (Witzmann, 2013). Convective heat exchange (C) between the skin and the environment is proportional to the difference between the skin and ambient air temperatures (Witzmann, 2013):

$$C = h_c \times A \times (T_{sk} - T_a) \quad [\text{W}] \tag{2.7}$$

Where h_c is the convective heat transfer coefficient (W/m^{2}K), A is the body surface area (m^2), T_{sk} and T_a are mean skin and ambient temperatures (K). The value of h_c includes the effects of the factors other than temperature and surface area that influence convective heat exchange, such as the air movement, direction of air movement and the curvature of the skin surface.

The ability of the human body to adjust the cutaneous vasomotor tone provides an effective means of modulating skin blood flow and therefore heat flux from core-to-skin. The cutaneous circulation acts as a thermoregulatory effector but is also affected by non-thermoregulatory responses. Two branches of the sympathetic nervous system, a noradrenergic active vasoconstrictor system and an active vasodilator system of uncertain neurotransmitter, control the cutaneous circulation. Sweating is the primary means of heat loss in humans (Witzmann, 2013).

The shell thermal conductance (corresponds to the convective heat transfer coefficient) is typically 5 to 9 W/m²K of the body surface if the skin blood flow is minimal. The heat produced in the body will be conducted to the surface of the body if there is the temperature difference between the core and the skin of at least 5 K. For a lean resting subject with a surface area of 1.8 m² with a minimal whole-body heat loss coefficient (also called the heat capacity rate) of 16 W/K and a metabolic heat production of 80 W (Hardy and DuBois, 1937; Witzmann, 2013).

The automatic regulation of heat loss was shown to be effective in environmental temperatures above 301.15 K (28°C), since the vasomotor changes combined with the activity of the sweat glands were capable of adjusting the loss of heat from the body into heat production. The thermal conductance of the body tissues remained constant with ambient temperatures below 301.15 K (28°C). In environments colder than 301.15 K (28°C), the body lost heat until the shivering mechanism was set off to increase the metabolism (Hardy, 1937). With the ambient temperatures from 303.15 to 305.15 K (30 to 32°C), the body is in a neutral zone and eliminates a minimal amount of heat, which is equal to the heat produced. The heat loss increased both in colder and warmer environments. The regulation of body heat loss in ambient temperatures above 303.15 K (30°C) is a combination of increased blood flow to the skin and by evaporation of the sweat. The sweating of the body increases with the increased blood flow to take care of the heat loss (Hardy and DuBois, 1937).

In a cool environment, the temperature of the skin will be low enough for this to occur, but with an ambient temperature of 306.15 K (33°C), the temperature of the skin is about 308.15 K (35°C), and without an increase in conductance, the core temperature would have to rise to 313.15 K (40°C) for the heat to be conducted to the skin. However, the increase in the skin blood flow increases the conductance of the shell and helps the body to re-establish thermal balance. Even a moderate rate of skin blood flow can have a dramatic effect on heat transfer. This is the effective way of the body to control dry (convective and radiative) heat loss by varying skin blood flow and the skin temperature. When there is no sweating, raising the skin blood flow brings skin temperature nearer to blood temperature, and lowering skin blood flow brings skin temperature nearer to ambient temperature – the skin blood flow and sweating work simultaneously to dissipate heat when the body is exercising. When exercising, there are further increases in skin blood flow, which causes only a slight change in skin temperature and dry heat exchange. This serves primarily to deliver the heat to the skin that is removed by the evaporation of sweat (Witzmann, 2013).

Since generally the heat flow is the amount of heat that is transferred per unit of time in the material, assuming conditions for maximum efficiency of heat transfer by the blood, the rate of heat flow (HF$_b$) as a result of convection by the blood is (Witzmann, 2013):

$$\dot{Q}_b = \mathrm{HF}_b = S_{\mathrm{kBF}} \times \left(T_c - \overline{T_{sk}}\right) \times 3.85 \text{ kJ/L°C} \quad [\mathrm{W} = \mathrm{J/s}] \tag{2.8}$$

Where HF$_b$ ($\dot{Q}_b$) is the rate of heat flow (W=J/s), S_{kBF} is the rate of the skin blood flow (L/s), T$_c$ is the core temperature (K), T$_{sk}$ is the temperature of the skin (K) and 3.85 kJ/L°C is the volume heat capacity, also termed the volume-specific heat of the blood. This equation is a reasonable approximation in a warm subject with moderate-to-high skin blood flow. In practice, the blood exchanges heat with the tissues through which it passes on its way to and from the skin and heat exchange with these other tissues is greatest when skin blood flow is low so the heat flow to the skin may be much less than predicted.

The heat capacity rate as a result of convection by the blood (C$_b$) is (Witzmann, 2013):

$$C_b = \mathrm{HF}_b \big/ \left(T_c - \overline{T_{sk}}\right) = S_{\mathrm{kBF}} \times 3{,}85 \text{ kJ/L°C} \quad [\mathrm{W/K}] \tag{2.9}$$

The total thermal heat capacity rate of the body shell is the sum of heat capacity rate T as a result of convection by the blood (C$_b$) and as a result of conduction of the heat through the tissues (C$_0$) (Witzmann, 2013):

$$\mathrm{HF} = (C_b - C_0) \times \left(T_c - \overline{T_{sk}}\right) \quad [\mathrm{W} = \mathrm{J/s}] \tag{2.10}$$

Where HF is the total rate of heat flow of the body (W=J/s), C$_b$ is the heat capacity rate as a result of convection by the blood (W/K), C$_0$ is the heat capacity rate of the tissues when skin blood flow is minimal (W/K), T$_c$ is the core temperature (K) and $\overline{T_{sk}}$ is the temperature of the skin (K). The maximum obtainable rate of heat flow (S_{kBF}) is estimated to be nearly 8 L/min. If the rate of blood flow is 1.89 L/min (0.0315 L/s), the skin blood flow contributes about 121 W/K to the conductance of the shell. If conduction through the tissues contributes 16 W/K, total shell heat capacity rate is 137 W/K, and if T$_c$=311.65 K (38.5°C) and T$_{sk}$=308.15 K (35°C) this will produce a core-to-skin heat transfer of 480 W (Witzmann, 2013).

The sweat glands are stimulated primarily by hypothalamic centres (usually parasympathetic). The sweat is secreted through the surface of the skin after the activation of the eccrine sweat glands, innervated by sympathetic cholinergic nerve fibres, which elicits the secretion of a precursor fluid. Sweat secretion is called a parasympathetic function, even though it is controlled by nerves distributed throughout the sympathetic nervous system. This is another means of promoting heat loss, by evaporation of the water content of the sweat. The eccrine glands cover most of the body (Wendt et al., 2007; Guyton and Hall, 2016).

Evaporation of the sweat (containing mostly the water) is an efficient way of losing heat, and it is the only means of losing heat from the body when the environment is warmer than 36°C, since then the environment is hotter than the skin. This avenue dissipates both heat produced by metabolic processes and any heat gained from the environment by convection and radiation. Most of the water evaporated in hot environments comes from sweat, but even in the cold the skin loses some water by the evaporation of insensible perspiration and diffusion through the skin rather than secretion (Witzmann, 2013).

The evaporative heat loss occurs independently of thermoregulatory control at the ambient temperature below 303.15 K (30°C). It accounts for 9 to 10 W/m² and corresponds to evaporation of about 13 to 15 g/(m²h), one-half of which is moisture lost through breathing, and the other half is lost through insensible perspiration. As the ambient temperature increases, the body depends more and more on the evaporation of sweat to achieve heat balance. The two histologic types of sweat glands are eccrine (the dominant type in all human populations with about 2.5 million.) and apocrine sweat glands. The eccrine sweat glands are more important in human thermoregulation, and they are controlled by cholinergic stimulation through postganglionic sympathetic C fibres that release acetylcholine (ACh) rather than norepinephrine. This elicits secretion of the precursor fluid. The apocrine sweat is more viscous and produced in much smaller amounts than eccrine sweat although the exact function of the apocrine glands is unclear, though they are thought to represent scent glands. A healthy man unacclimatised to heat can secrete up to 1.5 L/h of sweat, and the secretion capacity of the individual glands can change due to heat acclimatisation (Witzmann, 2013).

The evaporative heat loss (E) from the skin to the environment is proportional to the difference between the water vapour pressure at the skin surface and the water vapour pressure in the ambient air (Witzmann, 2013):

$$E = h_e \times A \times (p_{sk} - p_a) \tag{2.11}$$

where h_e is the evaporative heat transfer coefficient, A is the body surface area, p_{sk} is the water vapour pressure at the skin surface, and p_a is the ambient water vapour pressure. The value of h_e includes the effects of the air movement, direction of the air movement and the geometric factors. If the skin is completely wet, the water vapour pressure at the skin surface is the saturation water vapour pressure at the temperature of the skin, and evaporative heat loss has the maximum possible value (E_{max}). When the skin is not completely wet, it is impractical to measure p_{sk}, and the coefficient of skin wettedness (w) is calculated. The skin wettedness depends on hydration of the epidermis and the fraction of the skin surface that is wet (the balance between secretion and evaporation of sweat). If secretion exceeds evaporation, the sweat will start to accumulate increasing wettedness and evaporative heat loss. If evaporation exceeds secretion, the reverse occurs. If sweat rate exceeds E_{max}, once wettedness becomes 1, the excess sweat drips from the body because it cannot evaporate (Witzmann, 2013).

All thermal radiation is in the infrared range at ordinary tissue and environmental temperatures, but only the sun emits large amounts of radiation in the visible and near-infrared range, and colours of skin and clothing affect heat exchange only in sunlight or bright artificial light (Witzmann, 2013).

In humans, the respiratory heat exchange is usually relatively small (Witzmann, 2013).

The radiative heat exchange (R) is proportional to the difference between the fourth powers of the absolute temperatures of the skin (T_{sk}) and of the radiant environment (T_r), and to the emissivity of the skin (Witzmann, 2013):

$$R = h_r \times e_{sk} \times (T_{sk} - T_r) \quad \left[W/m^2 \right] \quad (2.12)$$

Where h_r is the radiative heat transfer coefficient (6.43 W/m²K) at 301.15 K (28°C), and e_{sk} is the emissivity of the skin.

Mechanisms of decreasing body temperature when it is too hot can be classified as (Guyton and Hall, 2016):

- *Vasodilation of skin blood vessels*: The vasodilatation of the skin blood vessels is triggered by the inhibition of the sympathetic centres (in the posterior hypothalamus that causes vasoconstriction) and can increase the rate of heat transfer to the skin as much as eightfold.
- *Sweating*: Sweating is triggered when the body core temperature rises above the critical level of 310.15 K (set point of 37°C) and induces the rate of evaporative heat loss. An increase of 1 degree in body temperature removes 10 times the basal rate of body heat production due to sweating.
- *Decrease in thermogenesis (heat production)*: The mechanisms of heat production (shivering and chemical thermogenesis) are inhibited.

2.3 THERMOPHYSIOLOGICAL ANALYSIS

When speaking of thermoregulation, one usually refers to four basic mechanisms called the heat loss and heat gain control mechanisms. These are sweating, shivering, vasodilatation and vasoconstriction. Sweating increases body heat loss by increasing sweat evaporation. Increased blood flow by vasoconstriction stimulates heat transfer from the body core to the skin by increasing conductance below the skin surface. Sweat serves to transfer the heat away from the surface of the body to the environment by means of convection and the evaporation. Vasodilatation and vasoconstriction are mechanisms that produce changes in the blood vessel diameter, thus affecting the rate of blood flow in the vessels and skin temperature.

Contrary to the above-described mechanisms of heat removal away from the body, the mechanisms of muscle tensing and shivering increase heat production and body temperature when cold. Shivering serves as a means of heat production, as it amplifies the involuntary movement of the muscle. Blood flow decreases due to vasoconstriction of the blood vessels and decreases conductance to preserve heat from escaping into the cold environment (Arens and Zhang, 2006).

2.3.1 The Human Thermal Homeostasis and Thermostatic Neural Mechanism

The basis for understanding human thermoregulatory responses through negative feedback mechanisms are traced as far back as the nineteenth century, when French physiologist Claude Bernard proposed the idea of active stabilisation of bodily states against disturbances from the outside. Bernard's work was later revived and crystallised in Walter B. Cannon's concept of homeostasis, which pointed out that the regulatory controls of the human body autonomously achieve stability and a condition of equilibrium within the body's economy (Cooper, 2008; Guyton and Hall, 2016).

In 1858, French physiologist Claude Bernard (1813–1878) introduced the term *internal environment* of warm-blooded organisms (Bernard, 1859; Davies, 2016). In 1865, his work was published for the American public under the title *An Introduction to the Study of Experimental Medicine* (Bernard, 1965). C. Bernard explained the concept by which only warm-blooded animals have the ability of the independent internal environment inside the organism, which is partially independent from the surrounding environment and able to maintain equilibrium by joining forces with these surrounding influences. He explained that warm-blooded animals, including humans, have the ability to maintain constant conditions within their own internal environment despite influences provided of the external environment. Humans are also warm-blooded animals, or homeotherms. His concept was based on the assumption that the internal environment will only become out of the balance if the organism's protective system fails to protect itself from external influences and becomes insufficient in given conditions. According to Claude Bernard, when discussing the warm-blooded animals, one has to understand that there are at least two environments to be considered – one being the external or extra-organic environment and other being the internal or intra-organic environment. The internal environment of living beings preserve the necessary relations of exchange and equilibrium with the external environment (Bernard, 1856; Gross, 1998). Claude Bernard moved from discussing the basic concept of animal heat and internal combustion to a discussion on thermoregulation by explaining that the nervous system controls regulatory functions and the circulatory system. He presented the sympathetic vasomotor control of neural regulation. The nervous system is accountable for controlling blood flows to the periphery of the body to facilitate cooling or to deeper tissue to counter heat loss, and thus is liable for maintaining body temperature in equilibrium (Cooper, 2008). He was the first to describe the third kind of thermoregulatory response, called *vasomotor action* and the first to acknowledge the control of the central nervous system over all thermoregulatory activities (Benzinger, 1965). American physiologist Walter Bradford Cannon (1871–1945) followed up on the ideas of Claude Bernard, and in 1926 W. B. Cannon derived the term *homeostasis*, which described the maintenance of a steady state in the human body and the physiological regulatory processes (Cannon, 1926, 1929). Cannon described the passive and active mechanisms of maintaining a steady state within an organism (Modell et al., 2015). W. B. Cannon explained homeostasis as the ability to maintain body functions within narrow limits. He also pointed to the complexity of the coordinated physiological processes, which maintain most of the steady states in the organism. He referred to these processes as conditions, which may vary but are relatively constant. Cannon proposed the usage of the term equilibria when describing the steady-state conditions, which the body tries to maintain (Cannon, 1932). He explained that living organisms have the ability to maintain constancy despite interference from the outer environment. He considered the human body as an open system, which always tries to establish and preserve a steady state and to keep balance. In this way, this open system requires mechanisms that act to maintain this constancy. To keep the body in a state of balance, the body resists changes induced by external circumstances or from within and produces the adaptive stabilising arrangements.

The concept of homeostasis by Walter B. Cannon includes (Cannon, 1932; Cooper, 2008):

- The human body is an open system and possesses mechanisms that act to maintain this constancy and a steady state.
- The steady-state concept is explained through the ability of the body to autonomously produce regulatory mechanisms that resist the change set by inner or outer effects.
- There are simultaneous cooperating mechanisms under the body's regulatory ability that maintains a steady state in the human body. The human regulatory system that determines a balanced state may be constituted of two or more antagonistic factors, which are brought into action at the same time or successively. Any factor that operates to maintain a steady state by an action in one direction does not act at the same point in the opposite direction. Factors that may be antagonistic in one region, where they effect a balance, may be cooperative in another region.
- Homeostasis is the consequence of gradual evolution and the result of organised self-governance.

A control system with homeostatic regulatory mechanisms under the rule of negative feedback systems operates in a way that causes any change to the regulated variable to be countered by a change in the effector output to restore the regulated variable towards its setpoint value (Modell et al., 2015).

The work of Claude Bernard and Walter B. Cannon opened the door to explaining human thermoregulatory mechanisms by introducing homeostasis as the maintenance of steady states in the body through coordinated physiologic mechanisms, some of which are certainly thermoregulatory mechanisms. The body has the capacity for self-regulation.

The homeostatic regulation of a physiologic variable often involves several cooperating mechanisms simultaneously. If the physiological variable is complicated, the more numerous and complicated mechanisms are activated to keep it at the desired value. Disease occurs if the body is unable to restore physiologic variables. The ability to maintain homeostatic mechanisms varies over a person's lifetime (Gallagher, 2013).

Today, modern physiologist's such as Guyton still quote Cannon's idea of homeostasis as the maintenance of nearly constant conditions in the internal environment (Modell et al., 2015). Different bodies control systems allow the functional systems to operate in support of one another. Homeostasis is maintained by a function of the control system, which initiates negative feedback, consisting of a series of changes that return the factor towards a certain mean value (Guyton and Hall, 2016).

The five critical components that a regulatory system must contain to maintain homeostasis (Modell et al., 2015) are:

- A sensor that measures the value of the regulated variable.
- A mechanism for establishing the 'normal range' of values for the regulated variable, also called the setpoint.
- An *error detector* that compares the signal being transmitted by the sensor (representing the actual value of the regulated variable) with the setpoint.
- The controller interprets the error signal and determines the value of the outputs of the effectors.
- The effectors are those elements that determine the value of the regulated variable.

The negative feedback control system consists of affiliated components, which control one another. These are a regulated variable, a sensor (or detector), a controller (or comparator) and an effector. If the disturbance within or outside the system starts to pressure the system, causing the change, the negative feedback mechanism will be activated. A sensor will detect the changes in the regulated variable, and the information will be sent back to the controller. Afterwards, the effector will be activated and will act to oppose change (Gallagher, 2013).

One example of a homeostatically regulated variable in the human body is the body core temperature (setpoint value is 309.15 to 310.65 K or 36 to 37.5°C and usually remains very constant, within ±0.6 K) with thermoreceptors (sensors) that detect the change. They send the information to the hypothalamus, which acts as a controller sending orders to effectors, which are the blood vessels, sweat glands in the skin and skeletal muscles. The effector response is a change in peripheral resistance, the rate of sweat secretion rate and shivering that serves to alter heat gains and losses (Modell et al., 2015; Guyton and Hall, 2016), Figure 2.1.

Most control systems of the body act through a negative feedback control mechanism (Guyton and Hall, 2016). Body temperature regulation works based on negative feedback. The sensors used to detect heat and cold are thermoreceptors. There are two kinds of mechanisms for human thermoregulation: chemical and physical temperature regulation (Guyton and Hall, 2016).

2.3.2 The Anatomy of Thermoregulation

In the 1970s the concept of the *set-point temperature* value (also called the reference or the equivalence or the feedback signal) was developed (Werner, 1979). James D. Hardy introduced the idea of the physiological thermostat of the human body, which has the characteristics of a proportional regulator with a setpoint (Hardy, 1965). Although some studies tried to oppose this idea and claimed there is a change in setpoint value during fever and temperature. Those studies postulated that if warm sensors become less responsive and cold sensors more responsive to temperature changes during fever, this will force a change in slope, which is indicative of a change in the gains of the thermoregulatory system during fever, rather than a change in the setpoint of the system (Mitchell et al., 1970; Cabanac et al., 1971;

Behind the Scenes

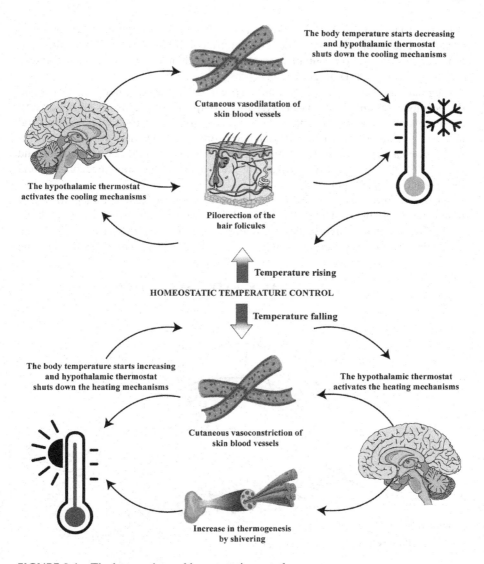

FIGURE 2.1 The human thermal homeostatic control.

Cabanac and Massonne, 1974). But this theory has never been accepted in physiology, so the negative feedback mechanism of the human thermoregulatory system was widely embraced.

The critical body core temperature is about 310.25 K (37.1°C) and this value is the *setpoint* of the temperature control mechanism. Temperature control mechanisms continually attempt to bring the body temperature back to this setpoint level. If the temperature rises above this value, the rate of heat loss is greater than that of heat production, forcing the body to decrease the internal core temperature closer to this value. If the body core temperature falls below this value, the rate of heat production is greater than that of heat loss, forcing the body to increase the internal core temperature. The feedback gain of the temperature control system is equal to the ratio of the change in environmental temperature to the change in body core temperature minus 1 (Guyton and Hall, 2016).

Regulation of the human body function is carried through the nervous system and the hormone system. The nervous system is composed of a sensory input (sensory receptors detect the state of the body or the state of the surroundings), the central nervous system (or integrative part that determines reactions in response to the sensations) and a motor output (transmits the signals to perform the action) (Guyton and Hall, 2016). The central nervous system (CNS) consists of the brain (encephalon) and the spinal cord, see Figure 2.2.

There are three major levels of the central nervous system with specific functional characteristics – the spinal cord, the lower brain or subcortical level (medulla, pons, mesencephalon, hypothalamus, thalamus, cerebellum and basal ganglia) and the higher brain or cortical level (Guyton and Hall, 2016). The CNS detects and

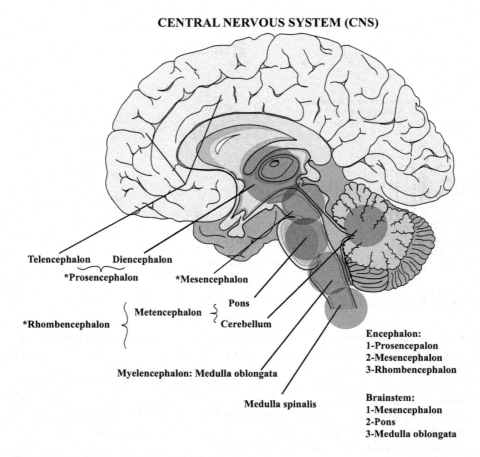

FIGURE 2.2 The organisation of the central nervous system.

processes sensory information from both the internal and external environment and comprises the following (Putz and Pabst, 2006; Kincaid, 2013b):

- Prosencephalon (also known as forebrain): telencephalon (also known as cerebrum consisting of basal nuclei and cerebral cortex) and diencephalon (composed of the thalamus, epithalamus, metathalamus, subthalamus and hypothalamus, with the third ventricle in the centre)
- Rhombencephalon (also known as hindbrain): metecephalon (consisting of the pons and cerebellum, also known as the little brain), myelencephalon (also known as medulla oblongata or afterbrain) and isthmus rhombencephali
- Mesencephalon (also known as midbrain)

The mesencephalon, metencephalon and the myelencephalon comprise the brain stem (also known as truncus encephali).

Many regions of the CNS control autonomic function through a hierarchy of reflexes and integrative centres (Kincaid, 2013a). Most of the subconscious activities of the body are controlled in the lower areas of the brain (Guyton and Hall, 2016). The centres located in the human brain, more specifically inside the hypothalamus, and operated by nervous feedback mechanisms (Manna, 2018) regulate the temperature of the human body (Manna, 2018).

The CNS directly controls the maintenance of body temperature (Rothwell, 1992). In other words, the hypothalamus is the part of the CNS that is related to monitoring and controlling the homeostatic function of the human body (Conn, 2017).

The hypothalamus has two-way communication pathways with all levels of the limbic system, sending output signals in three directions (Guyton and Hall, 2016):

- Backward and downward to the brain stem and afterwards into the peripheral nerves of the autonomic nervous system
- Upward towards many higher areas of the diencephalon and cerebrum
- Into the hypothalamic infundibulum (to control the secretory functions of pituitary glands)

The areas of the cerebral hemispheres, diencephalon, brainstem and central pathways to the spinal cord that are involved in the control of autonomic functions are collectively termed the central autonomic network – the areas of the hypothalamus control circadian rhythms and homeostatic functions, such as thermoregulation. Because of the major role of the hypothalamus in autonomic function, it is known as the head ganglion of the autonomic nervous system (ANS) (Kincaid, 2013a). The ANS is activated mainly by centres located in the spinal cord, brain stem and hypothalamus and controls most visceral functions of the body (Guyton and Hall, 2016). The ANS is usually subdivided into three divisions: sympathetic, parasympathetic and enteric. The ANS is an efferent system of the body, and the sensory neurons innervating the involuntary organs are not considered part of the ANS (Kincaid, 2013a). The efferent autonomic signals are transmitted to the various organs of the body through the sympathetic nervous system and the parasympathetic nervous system (Guyton and Hall, 2016).

The ANS regulates the involuntary functions of the body in order to regulate the body's internal environment. Parasympathetic postganglionic fibres release acetylcholine, whereas the sympathetic fibres release norepinephrine (Kincaid, 2013a). The terminals of neurons whose cell bodies are located in the brain stem and hypothalamus and most postganglionic neurons of the sympathetic nervous system secrete norepinephrine (Guyton and Hall, 2016). The fibres that secrete acetylcholine are cholinergic and oppose fibres that secrete norepinephrine, which are adrenergic (Guyton and Hall, 2016).

2.3.2.1 The Central Thermostatic Control and Hypothalamus

One of the most critical functions of the nervous system is the regulation of the body. The body core temperature is considered as a regulated variable in the thermoregulatory system. It is maintained by a combination of feedback and feedforward mechanisms. Feedback mechanisms are triggered when the core temperature deviates from the defended range, while the feedforward mechanisms are triggered in the absence of any change in core temperature. Although feedforward and feedback signals transfer different types of information, they converge on a common set of neural substrates in the preoptic area of the hypothalamus (Lek Tan and Knight, 2018).

The hypothalamus is the portion of the anterior end of the diencephalon that lies below the hypothalamic sulcus and in front of the interpeduncular nuclei (Manna, 2018), Figure 2.3. The hypothalamus, which represents less than 1% of the brain mass, is one of the most important of the control pathways of the limbic system and controls the internal functions and vegetative functions of the brain (Guyton and Hall, 2016), see Figure 2.3.

The vegetative and control functions of the hypothalamus (Guyton and Hall, 2016):

1. Cardiovascular regulation
2. Body temperature regulation
3. Body water regulation
4. Regulation of uterine contractility and milk ejection from the breasts
5. Gastrointestinal and feeding regulation
6. Hypothalamic control of endocrine hormone secretion by the anterior pituitary gland

The anterior portion of the hypothalamus, especially the preoptic area, is concerned with the regulation of body temperature (Guyton and Hall, 2016). The anterior hypothalamus has a central site for regulation of vasodilation and sweating and acts as a terminal sensory receptor organ for temperature independent of the heat stimulation of the skin. The posterior hypothalamus processes information and acts as the central site for regulating metabolism like a synaptic relay station for afferent impulses from receptors in the skin. This synaptic relay station is not affected by its own temperature, but it may be influenced by the anterior hypothalamus, which is responsive to its own temperature (Hammel et al., 1963).

The anterior hypothalamic-preoptic area is activated during the thermostatic detection of temperature and acts as thermostatic body temperature control centre,

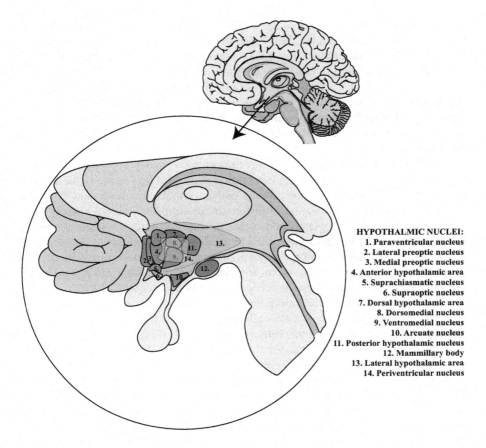

FIGURE 2.3 The hypothalamus.

while the posterior hypothalamus integrates the central and peripheral temperature sensory signals (Guyton and Hall, 2016). The preoptic-anterior hypothalamus senses changes in the body core temperature and also receives afferent sensory input from thermoreceptors throughout the body, including the spinal cord, abdominal viscera, the greater veins and the skin (Wendt et al., 2007; Witzmann, 2013).

To maintain core temperature within a narrow range (309.65 to 310.65 K or 36.5 to 37.5°C), the thermoregulatory system needs continuous information, which is received through the temperature-sensitive neurons and nerves (Witzmann, 2013). The relative contributions to the control signal from the hypothalamus (core) temperature and from the skin temperature are in the order of 10:1, so the system is heavily weighted towards representing the overall body thermal state (Arens and Zhang, 2006).

In the anterior hypothalamic-preoptic area there is a centre for detecting heat loss which was discovered by Aronsohn and Sachs in 1885. This heat loss centre in the anterior hypothalamic-preoptic area functions as a terminal temperature receptor organ with first neurons (Benzinger et al., 1961). It contains a large number of heat-sensitive neurons and around three times as many cold-sensitive neurons. The heat-sensitive neurons increase their firing rate 2- to 10-fold in response to a 10°C increase in body temperature. The cold-sensitive neurons increase their firing rate when the body temperature falls (Guyton and Hall, 2016).

This centre acts as a terminal receptor organ for temperature and controls both the direct responses to heat by vasodilatation and sweating through efferent pathways, and it indirectly controls, with counteracting impulses, the response by metabolic heat production to cold-stimulation of the skin through a third efferent pathway (Benzinger et al., 1961).

Three efferent pathways upon the three unconscious thermoregulatory functions (Benzinger et al., 1961) manage the innervation of the heat loss centre in the anterior hypothalamic-preoptic area. The efferent pathways that carry responses of sweating and peripheral vasodilatation, transfer stimuli from the internal sensory thermoreceptive site to sweat glands and cutaneous vessels. After the internal sensory thermoreceptive site has been innervated by the metabolic response to the increased frequency of cold impulses in the afferent pathway for cold-receptive stimuli, it is counteracted and depressed by warm impulses through another efferent pathway, originating from an internal thermoreceptive site (Benzinger et al., 1961).

When the anterior hypothalamic-preoptic area detects heat, it triggers reactions to cause the body to lose heat. The skin all over the body starts to sweat, and the skin blood vessels over the entire body dilate (Guyton and Hall, 2016).

In the posterior hypothalamus, there is a centre for maintaining heat. Simply put, this centre functions as a synaptic relay station to increase frequency from thermoreceptive endings in the skin for increasing the response of metabolic heat production. This centre is indifferent to the stimulus of temperature. It is stimulated by two afferent pathways, receiving stimuli, and transfers stimuli to the periphery by one efferent pathway. This centre receives stimuli from cold-receptive nerve endings of the skin through an afferent pathway to trigger the thermoregulatory heat production. After receiving the information on cold for chemical heat regulation from the skin as a terminal receptor site, this centre will initiate thermoregulatory responses through an efferent pathway and transfer stimuli to the metabolising tissues of the effector organs (Benzinger et al., 1961).

Only one pathway is efferent to the heat loss in the anterior hypothalamus but afferent to the centre for maintaining heat in the posterior hypothalamus. This efferent pathway is responsible for central depression and thereby controls shivering when the temperature increases at the heat loss centre in the anterior hypothalamus. The sensory impulses received from an internal receptor site for detecting heat loss in the anterior hypothalamus through the effector neurons of that pathway will trigger the depression of thermoregulatory heat production (Benzinger et al., 1961).

The temperature receptors that generate the signal for stimulation of the thermostatic control centre in the hypothalamus are located in the core of the body, directly inside the hypothalamus, and in other parts of the body, thermoreceptors in the skin and in a few specific deep tissues of the body (Guyton and Hall, 2016). So, the anatomy of the thermoregulation can be divided into the central thermostat located in the brain and peripheral sensing units.

The thermos-sensitive nerve endings, which serve as thermoreceptors, are located in the skin. The send stimulating signals through the sympathetic nerve system to the anterior hypothalamic-preoptic area. Afterwards, these signals can be sent to the posterior hypothalamus, which acts as a controller of body temperature during cold. There are cold-sensitive thermoreceptors in the anterior hypothalamus, in the spine and the abdomen. The posterior hypothalamus emits nerve signals to the periphery, stimulating vasoconstriction and shivering of the skin's blood vessels; it also initiates the release from the medulla of hormonal messengers such as norepinephrine that rapidly initiate vascular contraction throughout the body (Arens and Zhang, 2006).

Areas in which thermosensitivity has been detected are the preoptic area and anterior hypothalamus in the brain, the spinal cord and the skin, and the abdominal viscera and other areas have been suspected in muscle tissue or the veins draining muscle (Stolwijk, 1971). The visceral organs of the body are responsible for a large proportion of the metabolic heat produced (Ruben, 1995).

2.3.2.2 The Peripheral Reception and Skin

The skin is the first barrier between the human body and its environment that prevents the uncontrolled loss or gain of heat and water. Most of the heat generated in the human body is exchanged between the body and the environment at the skin's surface. In accordance with most of the heat being exchanged through the skin, the skin's temperature is much more variable than the core temperature. The temperature of the skin is affected by both internal thermoregulatory responses such as skin blood flow and sweat secretion, the temperatures of underlying tissues and outer environmental factors, such as the air temperature and movement and thermal radiation. Skin temperature is one of the major factors in heat exchange with the environment and provides the thermoregulatory system with important information about the need to conserve or dissipate heat (Witzmann, 2013).

The skin is about 2 mm thick, composed of the epidermis (0.075–0.15 mm thick) and dermis (much thicker than the epidermis), see Figure 2.4. The dermis contains complex vascular systems, sweat glands and thermoregulatory nerves. The changes in the vascular system due to vasoconstriction and vasodilatation control the skin's conductance. The thermos-sensitive nerve endings are located in the skin. They detect changes in the skin's temperature and transmit the information to the brain (Arens and Zhang, 2006).

The somatosensory system is a diverse sensory system comprising receptors and processing centres that produce sensory modalities, such as touch (tactile sensation), proprioception (body position), temperature and pain (nociception) (Bell and Rhoades, 2013).

Sensory information from the receptors, as well as the detection of warm and cold, is transmitted to the somatic portion of the sensory system. The thermoreceptive senses that detect heat and cold are classified as somatic senses, together with mechanoreceptive somatic senses and pain senses (Guyton and Hall, 2016). Information of thermal sensation enters the CNS through peripheral nerves and is conducted immediately to the ventrobasal complex of the thalamus and to the cerebral somatic sensory cortex (Guyton and Hall, 2016).

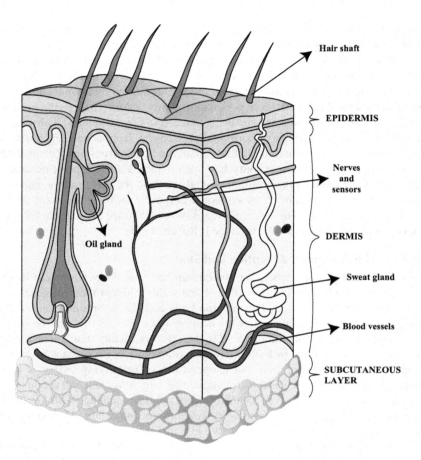

FIGURE 2.4 Cross-section of the skin.

There are three main types of peripheral sensory thermoreceptors, and they include cold receptors, warmth receptors and pain receptors. Primary afferent nerve fibres transduce, encode, and transmit thermal information and mediate the cutaneous thermosensation. They are located in the skin and in the oral and urogenital mucosa (Witzmann, 2013).

The thermoreceptors are located immediately under the skin at discrete separated spots. The cold-sensitive thermoreceptors are most numerous and most superficially situated. Usually, all skin areas have 3 to 10 times more cold spots than warmth spots. Their number varies from 15 to 25 cold spot/cm^2 in the lips to 3 to 5 cold spot/cm^2 in the fingers to less than 1 cold spot/cm^2 in some broad surface areas of the trunk. Humans are able to perceive different gradations of cold and heat, from freezing cold to cold, to cool to indifferent to warm sensations, from warm, to hot, to burning hot (Guyton and Hall, 2016). The cold sensors are closer to the surface than the warm, so these peripheral sensors are more dedicated to the rapid detection of cold than of warmth. The thermoreceptors are located in the dermis, with the cold receptors placed immediately beneath the epidermis and the warmth receptors placed within the upper layer of the dermis. The average depth for cold receptors is 0.15 to 0.17 mm, and 0.3 to 0.6 mm for warmth receptors (Arens and Zhang, 2006).

Cutaneous thermosensation is mediated by various primary afferent nerve fibres (Witzmann, 2013). Chemosensory fibres refer to thin and slow-conducting unmyelinated C fibres and thin and fast-conducting myelinated Aδ fibres (also called the trigeminal nerve fibres) that are involved in thermoreception (temperature perception) and nociception. They don't operate independently. The input from C cold fibres is inhibited by input from Aδ cool fibres (Conn, 2017).

The thermoreceptors are nonspecialised naked nerve endings supplied by chemosensory fibres (Bell and Rhoades, 2013). The warmth nerve endings are presumed to be a free nerve ending, while the cold receptors have been identified histologically with great certainty. The warmth signals are transmitted mainly over type C nerve fibres at transmission velocities of only 0.4 to 2 m/s. The cold signals are sensed and transmitted by type Aδ myelinated nerve endings, which branch into the bottom surfaces of basal epidermal cells. The cold signals are transmitted from Aδ myelinated nerve endings via type Aδ nerve fibres at velocities of about 20 m/s. Some cold sensations are also believed to be transmitted in type C nerve fibres, which suggests that some free nerve endings might also function as cold receptors (Guyton and Hall, 2016).

The mechanisms of peripheral thermoreceptor activation involves changes in the resting membrane potential since they are not pacemakers like the central thermosensitive neurons. The response of most peripheral thermoreceptors shows a powerful dynamic phasic component since these neurons are very active when detecting temperature changes but quickly adapt when the temperature stabilises, enabling rapid reaction to thermal environmental changes. Thermal transduction involves the action of transient receptor potential ion channels whose activity depends on the temperature of their environment.

There are three different types of the thermoreceptors – the vanilloid transient receptor potential ion channels, the melastatin (or long) transient receptor potential ion channels and the ankyrin transmembrane protein channels. Each of them operates over a specific temperature range, thereby providing a potential molecular basis for peripheral thermosensation. These specialised thermoreceptors are embedded in the membranes of afferent fibre terminals as free nerve endings in the skin. Sensing cold involves the cold-induced inhibition of K+ conductances, whereas sensing warmth may involve the otherwise mechanosensitive K2P channels and other nociceptive C fibres (Witzmann, 2013).

The cold and warmth receptors are stimulated by changes in their metabolic rates, which is the consequence of the temperature altering the rate of intracellular chemical reactions more than twofold for each 10-degree change. So, thermal detection is not the result of direct physical effects of heat or cold on the nerve endings but from chemical stimulation of the endings as modified by temperature (Guyton and Hall, 2016).

The thermal stimuli can even excite the pain receptors (besides the mechanical and the chemical stimuli). The mechanical or thermal types of stimuli generally stimulate sharp pain, also called the fast pain sensation, whereas all three types of stimuli can elicit a slow pain sensation, also called chronic pain. The fast pain sensation is felt within about 0.1 seconds after stimulation, contrary to slow pain sensation, which begins after 1 second or more and then increases slowly. The pain receptors, associated with thermal sensations, are stimulated only by extreme degrees of freezing cold and burning hot (Guyton and Hall, 2016).

The stimulation of the thermoreceptors shows the effects of different temperatures on the responses of types of nerve fibres and these fibres respond differently at different levels of temperature. The perception of temperature stimuli is closely related to the properties of the thermoreceptors. There are the pain fibres stimulated by cold, the cold fibres, the warmth fibres and the pain fibres stimulated by heat. The cold-pain fibres are stimulated due to extreme cold (lower than 290.15 K or lower than 17°C). As the temperature rises to 288.15 K (15°C), the cold receptors are stimulated, reaching peak stimulation at about 297.15 K (24°C) and fading out slightly above 313.15 K (40°C). In the comfort zone (from 303.15 to 309.15 K, or respectively from 30 to 36°C for a small area of skin), there is no appreciable temperature sensation. Above about 303.15 K (30°C), the warmth receptors are stimulated and fade out at about 322.15 K (49°C). The heat-pain fibres are stimulated at around 318.15 K (45°C), and also some of the cold fibres begin to be stimulated again, possibly because of damage to the cold endings caused by the excessive heat (Bell and Rhoades, 2013; Guyton and Hall, 2016).

There is a constant production of heat in the body's muscles, and the heat is transferred to all parts of the body by means of the bloodstream (Du Bois, 1939). Heat loss is greatly accelerated by cutaneous vasodilation and sweating after the metabolic heat has been transported to the skin (Wendt et al., 2007).

Vascular control is the consequence of the posterior hypothalamus or skin stimulation (Arens and Zhang, 2006). The blood flow in the skin, both cutaneous and subcutaneous, is controlled largely by the central nervous system through the sympathetic nerves (Guyton and Hall, 2016). The central nervous system, or precisely the vasomotor centre, controls the sympathetic vasoconstrictor and vasodilator system. The sympathetic impulses are transmitted to the adrenal medullae and cause the secretion of the epinephrine and norepinephrine, called the sympathetic vasoconstrictor neurotransmitters, into the circulating blood. These two hormones are carried in the bloodstream to all parts of the body, where they act directly on all blood vessels, usually to cause vasoconstriction. The anterior hypothalamus is the principal area of the brain controlling the vasodilator system (Guyton and Hall, 2016).

The thermal resistance of the epidermis and outer dermis determines the heat flow, while variations of blood flow within the small dermal capillaries are thermally insignificant. The deeper subcutaneous region contains the venous plexus affected by the vasoconstriction and vasodilatation, which strongly affects skin temperature and heat transfer from the skin to the environment. Blood flow into the venous plexus is fed by arterioles. Due to vasodilatation, the diameter of the blood vessels start to expand in the heat. Dilatation can cause an eightfold increase in skin conductance, producing a gradient from the body's central core temperature to skin surface temperature that is less than 1 K. A doubling in diameter corresponds to a 16-fold increase in the blood supply volume. When cold, blood vessels start to constrict, resulting in the low blood supply to the venous plexus as low as zero and a local gradient across the skin of 10 K (Arens and Zhang, 2006). The skin blood flow is around

3 mL/min/100 g of tissue in cool weather. But skin blood flow may increase to as high as 7 to 8 L/min for the entire body when humans are exposed to body heating (Guyton and Hall, 2016).

Humans can tolerate environmental temperatures considerably above the deep body temperature because of the ability to sweat. Due to sweating, a large quantity of heat can be dissipated due to evaporative loss, so that body temperature is regulated even under a relatively large heat load (Ingram and Mount, 1975). The vaporisation from the skin is the primary mechanism for losing excess body heat during activity, but evaporation can occur if the skin water vapour pressure (p_{sk}) is higher than the ambient air-water-vapour pressure (p_a). Air movement around the body also helps with sweat evaporation. When the air velocity increases, the boundary layer between the person and environment decreases thus allowing sweat evaporation (Li et al., 1995; Plante et al., 1995). A man can sweat profusely and lose heat at a rate equivalent to 20 times the resting metabolism (Ingram and Mount, 1975). Sweat accumulation within the clothing will not start during activity but when a person returns to resting state. The metabolic rate and sweating will then decrease, but the moisture will continue to evaporate and provide unwanted cooling (Huang, 2006).

The maximum sweat rate (SW_{max}) for non-acclimatised subjects as the function of the metabolic rate (M) can be calculated as (Malchaire et al., 2000):

$$SW_{max} = 2.62M - 149 \quad [g/h]. \tag{2.13}$$

The maximum sweat rate in non-acclimatised subjects falls in the range of 250 to 400 W/m² (equal to 650 to 1000 g/h).

The maximum sweat rate (SW_{max}) for acclimatised subjects as a function of the metabolic rate (M) can be calculated as (Malchaire et al., 2000):

$$SW_{max} = 3.27M - 186 \quad [g/h]. \tag{2.14}$$

The effectiveness of evaporative heat transfer depends on the vapour pressure difference between the respiratory/cutaneous surfaces and the environment (Ingram and Mount, 1975).

During moderate sweating, the vapour concentration does not exceed the saturated vapour concentration, and vapour concentration at the skin will increase to produce a vapour flow. This will enable the sweat to evaporate and heat to be carried away from the skin through the clothing and enable body cooling. When the sweat is unable to evaporate from the human skin due to increased vapour concentration, it will start to condensate and absorb in the clothing. The evaporated sweat will also condensate inside the clothing if the evaporation rate starts to increase (Lotens et al., 1990).

Evaporative heat loss in humans and animals becomes more important for maintaining body temperature as the ambient temperature rises. The vaporisation of water from the body may take place from the respiratory tract and from the surface of the skin. From the surface of the skin, the vaporisation is obtained by water lost by diffusion through the skin and not subject to physiological control, by sweating from special glands, which is under physiological control or by evaporation of moisture applied externally to the body surface (Ingram and Mount, 1975).

REFERENCES

Albuquerque Neto, C., L. F. Pellegrini, M. S. Ferreira, S. de Oliveira Jr. and J. I. Yanagihara. 2010. Exergy analysis of human respiration under physical activity. *International Journal of Thermodynamics* 13(3): 105–109.

ANSI/ASHRAE Standard 55-2017 Thermal Environmental Conditions for Human Occupancy. 2017. Atlanta, GA: American Society of Heating, Refrigerating and Air-Conditioning Engineers Inc. (ASHRAE), ISSN 1041-2336.

ASHRAE Handbook Fundamentals (SI ed.). 2017. Atlanta, GA: American Society of Heating, Refrigerating, and Air Conditioning Engineers Inc., ISBN: 1931862702.

Arens, E. A and H. Zhang. 2006. The skin's role in human thermoregulation and comfort. In: *Thermal and Moisture Transport in Fibrous Materials*, ed. N. Pan and P. Gibson. Cambridge, UK: Woodhead Publishing Ltd, 560–602. ISBN 978-1-84569-057-1.

Balmer, R. T. 2011. *Modern Engineering Thermodynamics*. Philadelphia, PA: Elsevier Inc., ISBN: 978-0-12-374996-3.

Bell, D. R. and R. A. Rhoades. 2013. Sensory physiology, Chapter 4. In: *Medical Physiology: Principles for Clinical Medicine*, 4th edition, ed. R. A. Rhoades and D. R. Bell. Philadelphia, PA: Lippincott Williams & Wilkins, ISBN: 9781609134273.

Benzinger, T. H. 1963. Peripheral cold- and central warm-reception, main origins of human thermal discomfort. *Proceedings of the National Academy of Sciences of the United States of America* 49: 832–839.

Benzinger, T. H. 1964. The thermal homeostasis of man. *Symposia of the Society for Experimental Biology* 18:49–80.

Benzinger, T. H. 1965. The thermal homeostasis of man. In: *Les concepts de Claude Bernard sur le milieu intérieur. An international symposium.* ed. R. Heim. Paris, France: Masson & C Editeurs, Libraires de L'Académie de Médicine, 326–379.

Benzinger, T. H., A. W. Pratt and C. Kitzinger. 1961. The thermostatic control of human metabolic heat production. *Proceedings of the National Academy of Sciences of the United States of America* 47(5): 730–739.

Bernard, C. 1856. *Leçons sur la physiologie et la pathologie du système nerveux. Leçon d'ouverture*. Volume 1: Cours de pathologie expérimentale. Paris, France. Republished in 1860 in The medical times.

Bernard, C. 1859. *Leçons sur les physiologiques et les altérations pathologiques des liquides de l'organisme*. Volume 2: Cours de médecine du Collège de France. Baillière, Paris, France.

Bernard, C. 1965. *An Introduction to the Study of Experimental Medicine*. New York, NY: Henry Schuman.

Boron, W. F. and E. L. Boulpaep. 2012. *Medical Physiology: A Cellular and Molecular Approach*, 2nd edition. Philadelphia, PA: Saunders, Elsevier Inc., ISBN: 978-1-4377-1753-2.

Butera, F. M. 1998. Principles of thermal comfort, Chapter 3. *Renewable & Sustainable Energy Reviews* 2(1–2): 39–66.

Buyak, N. A., V. I. Deshko and I. O. Sukhodub. 2017. Buildings energy use and human thermal comfort according to energy and exergy approach. *Energy and Buildings* doi:10.1016/j.enbuild.2017.04.008.

Cabanac, M. and B. Massonne. 1974. Temperature regulation during fever: Change of set point or change of gain? A tentative answer from a behavioural study in man. *Journal of Physiology* 238: 561–568.

Cabanac, M., D. J. C. Cunningham and J. A. J. Stolwijk. 1971. Thermoregulatory set point during exercise: A behavioral approach. *Journal of comparative and physiological psychology* 76: 94–102, American Psychological Association.

Caliskan, H. 2013. Energetic and exergetic comparison of the human body for the summer season. *Energy Conversion and Management* 76: 169–176.
Canadas, V. 2005. To be or not to be comfortable: Basis and prediction. In: *Environmental Ergonomics: The Ergonomics of Human Comfort, Health and Performance in the Thermal Environment*, 1st edition, ed. Y. Tochihara and T. Ohanaka. London, UK: Elsevier Ltd. ISBN: 978-0-0804-4466-0.
Cannon. W. B. 1926. Physiological regulation of normal states: Some tentative postulates concerning biological homeostatics. In: *À Charles Richet: ses amis, ses collègues, ses élèves*. Paris, France: Éditions Médicales, 91–93.
Cannon, W. B. 1929. Organization for physiological homeostasis. *Physiological Reviews* 9: 399–431.
Cannon. W. B. 1932. *The Wisdom of the Body*. New York, NY: W. W. Norton & Company, 177–201.
Charkoudian. N. 2016. Human thermoregulation from the autonomic perspective. *Autonomic Neuroscience: Basic and Clinical*. doi:10.1016/j.autneu.2016.02.007.
Cheng, Y., J. Niu and N. Gao. 2012. Thermal comfort models: A review and numerical investigation. *Building and Environment* 47: 13–22.
Choi, W., R. Ooka and M. Shukuya. 2018. Exergy analysis for unsteady-state heat conduction. *International Journal of Heat and Mass Transfer* 116: 1124–1142.
Conn, P. M. 2017. *Conn's Translational Neuroscience*. London, UK: Elsevier Inc., ISBN: 978-0-12-802381-5.
Cooper, S. J. 2008. From Claude Bernard to Walter Cannon. Emergence of the concept of homeostasis. *Appetite* 51: 419–427.
Cramer, M. N. and O. Jay. 2016. Biophysical aspects of human thermoregulation during heat stress. *Autonomic Neuroscience: Basic and Clinical*. doi:10.1016/j.autneu.2016.02.007.
d'Ambrosio Alfano, F. R., B. W. Olesen and B. I. Palella. 2017. Povl Ole Fanger's impact ten years later. *Energy and Buildings*. doi:10.1016/j.enbuild.2017.07.052.
Davies, K. J. A. 2016. Adaptive homeostasis. *Molecular Aspects of Medicine*. doi:10.1016/j.mam.2016.04.007.
Diller, K. R., J. W. Valvano and J. A. Pearce. 2000. Bioheat transfer, Chapter 4.4. In: *The CRC Handbook of Thermal Engineering*, ed. F. Kreith. Boca Raton, FL: CRC Press, ISBN: 0-8493-9581-X.
Dincer, I. and M. A. Rosen. 2007. *Exergy: Energy, Environment And Sustainable Development*, 1st edition. Philadelphia, PA: Elsevier Inc., ISBN: 978-0-0804-4529-8.
Dovjak, M., M. Shukuya and A. Krainer. 2015. Connective thinking on building envelope – Human body exergy analysis. *International Journal of Heat and Mass Transfer* 90: 1015–1025.
Du, C., B. Li, Y. Cheng, C. Li, H. Liu and R. Yao. 2018. Influence of human thermal adaptation and its development on human thermal responses to warm environments. *Building and Environment*. doi:10.1016/j.buildenv.2018.05.025.
Dubois, D. and E. F. Dubois. 1916. A formula to estimate the approximate surface area if height and weight be known. *Archive of Internal Medicine* 17 (6,2): 863–871.
DuBois, E. F. 1939. Heat loss from the human body. Harvey Lecture. *The Bulletin* 143–173.
Fanger, P. O. 1970. *Thermal Comfort: Analysis and Applications in Environmental Engineering*. Copenhagen, Denmark: Danish Technical Press.
Fanger, P. O. 1973a. *Thermal Comfort: Analysis and Applications in Environmental Engineering*. New York, NY: McGraw-Hill Book Company, ISBN: 0-07-019915-9.
Fanger, P. O. 1973b. Assessment of man's thermal comfort in practice. *British Journal of Industrial Medicine* 30: 313–324.
Fiala, D., K. J. Lomas and Martin Stohrer. 1999. A computer model of human thermoregulation for a wide range of environmental conditions: The passive system. *Journal of Applied Physiology* 87: 1957–1972.

Gagge, A. P. and Y. Nishi. 1977. Heat exchange between human skin surface and thermal environment, Chapter 5. In: *Handbook of Physiology, Supplement 26: Reactions to Environmental Agents*, ed. D. H. K. Lee, H. L. Falk and S. D. Murphy, 69–92. Rockville, MD: American Physiological Society, ISBN: 9780470650714.

Gallagher, P. J. 2013. Homeostasis and cellular signalling, Chapter 1. In: *Medical Physiology: Principles for Clinical Medicine*, 4th edition, ed. R. A. Rhoades and D. R. Bell. Philadelphia, PA: Lippincott Williams & Wilkins, ISBN: 9781609134273.

Gross, C. G. 1998. Claude Bernard and the constancy of the internal environment. *Neuroscientist* 4: 380–385.

Guyton, A. C. and J. E. Hall. 2016. Guyton and Hall textbook of medical physiology, 13th edition. Philadelphia, PA: Elsevier Inc., ISBN: 978-1-4557-7005-2.

Hammel, H. T., D. C. Jackson, J. A. J. Stolwijk, J. D. Hardy and S. B. StrØmme. 1963. Temperature regulation by hypothalamic proportional control with an adjustable set point. *Journal of Applied Physiology* 18(6):1146–1154.

Hardy, J. D. 1937. The physical laws of heat loss from the human body. *Proceedings of the National Academy of Sciences of the United States of America* 23 (12): 631–637.

Hardy, J. D. 1965. The set point concept. In: *Physiological Controls and Regulations*, ed. W. S. Yamamoto and J. R. Brobeck. Philadelphia, PA: Saunders, 98–116.

Hardy J. D. and E. F. DuBois. 1937. Regulation of heat loss from the human body. *Proceedings of the National Academy of Sciences of the United States of America* 23(12): 624–631.

Havenith, G. 1999. Heat balance when wearing protective clothing. *Annals of Occupational Hygiene* 43(5): 289–296.

Henriques, I. B., C. E. K. Mady and S. de Oliveira Jr. 2017. Assessment of thermal comfort conditions during physical exercise by means of exergy analysis. *Energy* 128: 609e617.

Holmér, I., P-O. Granberg and G. Dahlström. 1998. Cold environment and cold work. In: Heat and cold, Chapter 42, *ILO Encyclopedia of Occupational Health and Safety*, Vol. 2, ed. J. J. Vogt. http://www.iloencyclopaedia.org/part-vi-16255/heat-and-cold/42/cold-environment-and-cold-work (accessed on August 2, 2018).

Huang, J. 2006. Thermal parameters for assessing thermal properties of clothing. *Journal of Thermal Biology* 31: 461–466.

Hulbert, A. J. and P. L. Else. 2000. Mechanisms underlying the cost of living in animals. *Annual Review of Physiology* 62: 207–235.

Ingram, D. L. and L. E. Mount. 1975 *Man and Animals in Hot Environments. Topics in Environmental Physiology and Medicine.* New York, NY: Springer-Verlag New York Inc., ISBN-13: 978-1-4613-9370-2.

ISO 7730:2005 Ergonomics of the Thermal Environment – Analytical Determination and Interpretation of Thermal Comfort Using Calculation of the PMV and PPD Indices and Local Thermal Comfort Criteria. 2005. ISO-International Organization for Standardization.

Ivanov, K. P. 2006. The development of the concepts of homeothermy and thermoregulation. *Journal of Thermal Biology* 31: 24–29.

Juusela, M. A. and M. Shukuya. 2014. Human body exergy consumption and thermal comfort of an office worker in typical and extreme weather conditions in Finland. *Energy and Buildings* 76: 249–257.

Kincaid, J. C. 2013a. Autonomic nervous system, Chapter 6. In: *Medical Physiology: Principles for Clinical Medicine*, 4th edition, ed. R. A. Rhoades and D. R. Bell. Philadelphia, PA: Lippincott Williams & Wilkins, ISBN 9781609134273.

Kincaid, J. C. 2013b. Integrative functions of the central nervous system, Chapter 7. In: *Medical Physiology: Principles for Clinical Medicine*, 4th edition, ed. R. A. Rhoades and D. R. Bell. Philadelphia, PA: Lippincott Williams & Wilkins, ISBN 9781609134273.

Kurz, A. 2008. Physiology of thermoregulation. *Best Practice & Research Clinical Anaesthesiology* 22(4): 627–644.

Lek Tan, C. and Z. A. Knight. 2018. Regulation of body temperature by the nervous system. *Neuron* 98: 31–48.
Li, Y., A. M. Plante and B. V. Holcombe. 1995. Fiber hygroscopicity and perceptions of dampness. Part 2. Physical-mechanisms. *Textile Research Journal* 65: 316–324.
Lotens, W. A., F. J. G. van de Linde and G. Havenith. 1990. *The Effect of Condensation in Clothing on Heat Transfer*, TNO report. Netherlands: TNO Institute for Perception.
Luo, M., Z. Wang, K. Ke, B. Cao, Y. Zhai and X. Zhou. 2018. Human metabolic rate and thermal comfort in buildings: The problem and challenge. *Building and Environment* 1–21.
Luo, M., Z. Wang, X. Zhou, Y. Zhu and J. Sundella. 2016. Revisiting an overlooked parameter in thermal comfort studies, the metabolic rate. *Energy and Buildings* 118: 152–159.
Mady, C. E. K., C. Albuquerque, T. L. Fernandes, A. J. Hernandez, P. H. N. Saldiva, J. I. Yanagihara and S. de Oliveira Jr. 2013. Exergy performance of human body under physical activities. *Energy* 62: 370–378.
Mady, C. E. K., M. S. Ferreira, J. I. Yanagihara and S. de Oliveira Jr. 2014. Human body exergy analysis and the assessment of thermal comfort conditions. *International Journal of Heat and Mass Transfer* 77: 577–584.
Mady, C. E. K., M. S. Ferreira, J. I. Yanagihara, P. H. N. Saldiva and S.de Oliveira Jr. 2012. Modelling the exergy behavior of human body. *Energy* 45: 546–553.
Malchaire, J., B. Kampmann, G. Havenith, P. Mehnert and H. J. Gebhardt. 2000. Criteria for estimating acceptable exposure times in hot working environments: A review. *International Archives of Occupational and Environmental Health* 73: 215–220.
Manna, S. 2018. *Review of Physiology*, 2nd edition. London, UK: Jaypee Brothers Medical Ltd., ISBN: 978-93-86322-1.
Mitchell, D., J. W. Snellen and A. R. Atkins. 1970. Thermoregulation during fever: Change of set point or change of gain. *Pflügers Archiv – European Journal of Physiology* 321: 293–302.
Modell, H., W. Cliff, J. Michael, J. McFarland, M. P. Wenderoth and A. Wright. 2015. A physiologist's view of homeostasis. *Advances in Physiology Education* 39: 259–266.
Moran, M. J. 2000. Exergy analysis, Chapter 1.5. In: *The CRC Handbook of Thermal Engineering*, ed. F. Kreith. Boca Raton, FL: CRC Press LLC, ISBN: 0-8493-9581-X.
Morrison, S. F. 2016. Central neural control of thermoregulation and brown adipose tissue. *Autonomic Neuroscience: Basic and Clinical* 1–11.
Orosa, J. A. 2009. Research on the origins of thermal comfort. *European Journal of Scientific Research* 34(4): 561–567.
Parsons, K. C. 1998. Assessment of heat stress and heat stress indices. In: Heat and cold, Chapter 42, *ILO Encyclopedia of Occupational Health and Safety*, Vol. 2, ed. J. J. Vogt. Geneva, Switzerland: International Labour Office. http://www.iloencyclopaedia.org/part-vi-16255/heat-and-cold/42/assessment-of-heat-stress-and-heat-stress-indicies (accessed on July 20, 2018).
Plante, A. M., B. V. Holcombe and L. G. Stephens. 1995. Fiber hygroscopicity and perceptions of dampness. Part 1. Subjective trials. *Textile Research Journal* 65: 293–298.
Prek, M. 2005. Thermodynamic analysis of human heat and mass transfer and their impact on thermal comfort. *International Journal of Heat and Mass Transfer* 48:731–739.
Prek, M. 2006. Thermodynamical analysis of human thermal comfort. *Energy* 31: 732–743.
Prek, M. and V. Butala. 2010. Principles of exergy analysis of human heat and mass exchange with the indoor environment. *International Journal of Heat and Mass Transfer* 53: 5806–5814.
Prek M. and V. Butala. 2017. Comparison between Fanger's thermal comfort model and human exergy loss. *Energy*. doi:10.1016/j.energy.2017.07.045.
Prek, M., M. Mazej and V. Butala. 2008. An approach to exergy analysis of human physiological response to indoor conditions and perceived thermal comfort. In: *7th International Thermal Manikin and Modelling Meeting*. University of Coimbra, Spain.

Proceedings of the 29th International Conference on Efficiency, Cost, Optimisation, Simulation and Environmental Impact of Energy Systems. 2016. ed. A. Kitanovski and A. Poredoš. Ljubljana, SLO: University of Ljubljana, Faculty of Mechanical Engineering, ISBN 978-961-6980-15-9.

Putz, R. and R. Pabst. 2006. *Sobotta Atlas of Human Anatomy, Volume 1: Head, Neck, Upper Limb*, 14th edition. Elsevier GmbH, ISBN 978-0-443-10348-3.

Rant, Z. 2001. *Termodinamika: knjiga za uk in prakso*. Ljubljana, SLO: University of Ljubljana, Faculty of Mechanical Engineering, ISBN 961-6238-47-7.

Rothwell, N. J. 1992. Eicosanoids, Thermogenesis and Thermoregulation. *Prostaglandins Leukotrienes and Essential Fatty Acids* 46 (1): 1–7.

Ruben, J. A. 1995. The evolution of endothermy in mammals and birds: From physiology to fossils. *Annual Review of Physiology* 57: 69–95.

Schweiker, M. and M. Shukuya. 2012. Adaptive comfort from the viewpoint of human body exergy consumption. *Building and Environment* 51: 351–360.

Sessler D. I. 2016. Perioperative thermoregulation and heat balance, *Lancet* 387: 2655–2664.

Shapiro Y. and Y. Epstein. 1984. Environmental physiology and indoor climate thermoregulation and thermal comfort. *Energy and Buildings* 7: 29–34.

Simone, A., J. Kolarik, T. Iwamatsu, H. Asada, M. Dovjak, L. Schellen, M. Shukuya and B. W. Olesen. 2011. A relation between calculated human body exergy consumption rate and subjectively assessed thermal sensation. *Energy and Buildings* 43: 1–9.

Stolwijk, J. A. J. 1971. *A Mathematical Model of Physiological Temperature Regulation in Man, NASA CR-1855*. New Haven, CT: John B. Pierce Foundation Laboratory and Department of Epidemiology and Public Health, Yale University School of Medicine, 1–77.

Wendt, D., L. J. C. van Loon and W. D. van Marken Lichtenbelt. 2007. Thermoregulation during exercise in the heat strategies for maintaining health and performance. *Sports Medicine* 37(8): 669–682.

Werner, J. 1979. The concept of regulation for human body temperature. *Journal of Thermal Biology* 5: 75–82.

Witzmann, F. A. 2013. Regulation of body temperature, Chapter 28. In: *Medical Physiology: Principles for Clinical Medicine*, 4th edition, ed. R. A. Rhoades and D. R. Bell. Philadelphia, PA: Lippincott Williams & Wilkins, ISBN 9781609134273.

Wua, X., J. Zhao, B.W. Olesen and L. Fang. 2013. A novel human body exergy consumption formula to determine indoor thermal conditions for optimal human performance in office buildings. *Energy and Buildings* 56: 48–55.

3 Modelling Heat Losses from the Human Body

The two possible modes of heat dissipation are sensible and evaporative (or latent) heat transfer. The term sensible indicates a mode of heat transfer that depends on a temperature gradient (radiation, convection and conduction), while the evaporative heat transfer does not necessarily depend on a temperature gradient but instead on the heat that is taken up by water when it changes from a liquid into a vapour state. Its particular significance is that loss of heat can still take place by this means even when the surroundings are at a higher temperature than the observed body (Ingram and Mount, 1975).

Heat transfer from the core to the cutaneous and respiratory surfaces is fully sensible, while the heat transfer from the surfaces to the environment is both sensible and evaporative. Sensible and evaporative heat transfers are in parallel and additive. Their magnitudes may be calculated using sensible and evaporative transfer coefficients, but they do, however, have different dimensions, because the sensible coefficient involves temperature and the evaporative coefficient involves vapour pressure (Ingram and Mount, 1975).

The evaporation of water is the only means of cooling when the temperature of the surroundings is above that of the body, while non-evaporative heat transfer is merely an added heat load. Evaporative heat loss increases under hot conditions, both from the skin surface and from the upper respiratory tract. An increased evaporative loss under hot conditions in humans takes place primarily from the skin. In a cool environment, evaporative heat loss accounts for about 25% of the heat produced by a resting metabolism in man, while under hot conditions the evaporative heat loss accounts for the total heat loss (Ingram and Mount, 1975).

The energy expenditure of the human subject is affected by their clothing's thermal insulation in all environmental conditions, especially during exposure to mild cold above 288.15 K (15°C), which does not induce visible shivering, and also in cold conditions below 288.15 K (15°C) (Shephard, 1987; Vogelaere et al., 1992; Lee and Choi, 2004).

The first studies on heat transfer within the human body–clothing–environment system involved applying the basic laws of physics as the starting point for explaining different modes of heat and mass transfer within this system.

Some of the known principles in physics applied to thermal comfort research studies are:

1. *Einstein's law of energy conservation* has been used to explain the total energy of system conservation during all of the interactions within the system. When the heat of the system changes, it will produce a change in molecular activity. In other words, mass change will always be accompanied

by a simultaneous change in kinetic energy, but the total energy of the system will remain constant.
2. *Newton's law of cooling* has been used to explain that dry heat loss is proportional to the temperature difference between the human body and its surroundings. This principle has been used to explain steady-state heat exchange, where the evaluation of the thermal conductance (C) is determined by standard metabolic rates at a variety of air temperatures. However, further studies proposed an adjustment of the thermal conductance in accordance with the air velocity effect (Hardy, 1937; DuBois, 1939; Hardy and Soderstrom, 1938; Tracy, 1972). The rate of moisture convection can also be expressed using a form of Newton's law of cooling (Bhatia and Malhotra, 2016).
3. *Stefan–Boltzmann and Planck's law* has been used to explain the radiative heat exchange within the system. Radiative heat transfer occurs via electromagnetic waves and does not require a medium. The energy radiating from an object and received by the human body is proportional to the temperature difference between the object and the skin. The energy radiated from an object is proportional to the fourth power of its Kelvin temperature and the higher the temperature the more it radiates. Surface radiation is controlled by surface temperature according to the Stefan–Boltzmann law, while the rate of convection heat transfer is controlled by the temperature gradient (DuBois, 1939; Prek, 2006).
4. *Fick's law and Gibbs–Dalton law of diffusion* have been used to explain heat loss via the diffusion of water vapour through the clothing. Applying Fick's law to textile materials, the water vapour flux per unit area is calculated as the difference between the water vapour concentration at the surface of the specimen (g/m^3) and the water vapour concentration of the ambient air (g/m^3), divided by the total resistance to water vapour diffusion (Nicholson et al., 1999; Huang, 2016; *ASHRAE Handbook of Fundamentals*, 2017). Mass flow with time depends on density and wind speed, but Fick's law can be written in terms of a partial pressure gradient instead of a concentration gradient for cases of dilute mass diffusion. If the diffusion of water vapour is due to the concentration gradient, it is assumed that local equilibrium exists through the gas mixture, that the gases are ideal and that the Gibbs–Dalton law is valid, which implies that the temperature gradient has a negligible effect (*ASHRAE Handbook of Fundamentals*, 2017).
5. *Lewis relation* is used to explain the dynamic vapour resistance of clothing (Parsons et al., 1999).
6. *Fourier's law* has been used to explain heat transfer due to conduction. This law in its basic form explains steady-state heat conduction in one dimension (*ASHRAE Handbook of Fundamentals*, 2017).

As seen in Chapter 2, both thermodynamical, physical and physiological approaches base their explanation on the existence of heat balance within this system. The thermodynamical approach explains thermal comfort in terms of the first law of thermodynamics and the ability of the human body to exchange energy with the environment as an open system. In contrast, the physiological approach corresponds to the ability

Modelling Heat Losses from the Human Body 45

of the body to voluntarily or involuntarily maintain the heat balance with its environment by the subconscious mechanisms of body temperature control, which is kept within a narrow range, and the behavioural control of temperature mechanism such as adaptation through clothing. This will be fully addressed in the following chapters.

3.1 THE COMFORT IN HUMANS

Human subjects might perceive thermal comfort differently even when exposed to the same thermal environment. Humans show an extended range of individual differences, which affect thermal comfort and should be carefully considered (Wang et al., 2018).

The term *comfort* is hard to define since it involves thermal and non-thermal components and it depends on different conditions. Comfort, in general, involves both the psychological, physiological and physical aspects influencing the human body, and is much affected by the subjective sensations.

The aspects of clothing comfort are (Slater, 1985; Li, 2001; Bartels, 2006):

1. Thermophysiological comfort:
   Thermophysiological comfort is the ability to attain a comfortable thermal and wetness state, or the ability to maintain thermal balance between the heat generated by the body and the heat lost, due to the heat and the moisture transport through fabrics and clothing. Thus, thermophysiological comfort is directly related to the role of textiles as a barrier between the human body and the environment. It may also be defined as the ability to maintain thermal balance between the heat generated by the body and the heat lost.
2. Sensory, also called sensorial, comfort:
   Sensorial comfort is related to the mechanical, thermal and visual sensations provided during interactions with the environment and the clothing.
3. Wearing comfort, also called the body movement or kinetic comfort:
   This type of ergonomic wear comfort is related to the fit of the clothing and the freedom of movement.
4. Aesthetic or psychological comfort:
   Body movement comfort is the ability of a textile to allow freedom of movement, reduced burden and body shaping and subjective comfort is related to the perception of the clothing, which contributes to the overall well-being of the wearer (Slater, 1985; Hatch, 1993; Angelova, 2016). Comfort is a psychological feeling or judgement of a wearer who wears the clothing under certain environmental conditions (Li, 2001). A modern interpretation states that thermal comfort is a thermal sensation of feeling comfortable (Rohles et al., 1973).

Comfort can be divided into (Slater, 1985):

1. Physiological comfort
2. Psychological comfort (involves the sensorial, tactile and physical factors)
3. Physical comfort

It is impossible to achieve all aspects of comfort, but efforts can be made in order to achieve an optimal combined comfort state (Kamalha et al., 2013). According to Slater, the physiological comfort is related to the human body's ability to maintain life. The physiological definition of comfort is defined as freedom from pain and from discomfort as a neutral state (Hatch, 1993). Psychological comfort is the ability of the mind to keep the body functioning without external help. Physical comfort involves the effect of the external environment on the body (Slater, 1985). Comfort involves thermal and non-thermal components related to wear situations such as working, non-critical and critical conditions (Fourt and Hollies, 1970). Sensorial comfort includes the various sensations due to whole or partial contact between the clothing and the wearer's skin (Kaplan and Okur, 2008). The combination of different influences provided by the fibres, yarns, fabrics, finishes and the wearer affect the overall sensorial comfort of the clothing (Behery, 2005).

The human body's physiological responses are predictable for a given combination of clothing and environmental conditions when the system reaches a steady state (Fourt and Hollies, 1970). According to the World Health Organisation (WHO), thermal comfort is about ensuring a sensation of satisfaction within the ambient temperature as well as other health-wise implications (http://www.euro.who.int). The WHO defines health as 'a state of complete physical, mental and social well-being and not merely the absence of disease or infirmity' (Fanger, 1970).

The well-known definition of thermal comfort is given in ASHRAE Standard 55-2017 (ANSI/ASHRAE Standard 55, 2017) and the *ASHRAE Handbook of Fundamentals* (*ASHRAE Handbook of Fundamentals*, 2017). According to this definition, based on the work of Powl Ole Fanger, thermal comfort is defined as a condition of the mind that expresses satisfaction with the thermal environment (Fanger, 1970). The thermal comfort zone is not a unified strict set of conditions, which should be provided, rather it is the span of conditions where 80% of sedentary or slightly active persons find the environment thermally acceptable (*ASHRAE Thermal Environmental Conditions for Human Occupancy*, 1992), Figure 3.1.

Clothing comfort is one of the most important attributes for this (Wong and Li, 2002), and the variety of factors influencing comfort in humans have been defined as (Li, 2001; Kamalha et al., 2013):

1. ***Physical factors***: fibre type, composition and structure, wicking, air permeability, thermal insulation, tensile, bending, bending/flexibility, shearing, compression, thickness and drapeability of fabrics, fabric thermal resistance, fabric moisture permeability or resistance, air permeability/breathability (fabric handle)
2. ***Psychophysical or neurophysiological factors*** (relate to the sensorial comfort and responses to physical stimuli like visual, thermal, pressure and tactile stimuli): tactile sensations like appearance, smoothness, roughness, prickliness, stickiness, scratchiness, softness, stiffness, coarseness and bulkiness; thermal sensations like warmth, coolness, breathability, hotness and chilliness, localised heating or cooling and draft sensation due to air

Modelling Heat Losses from the Human Body

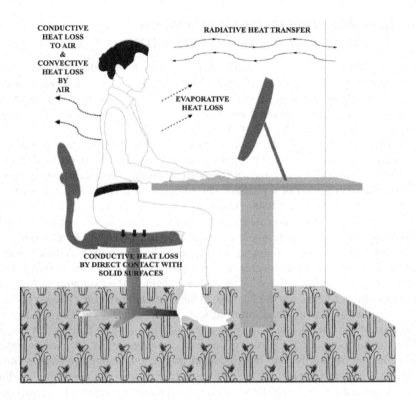

FIGURE 3.1 The heat loss mechanisms in a sedentary person.

movement; moisture sensations like clamminess, dampness or dryness, wetness, stickiness, non-absorbency and clinginess; pressure sensations like fit, snugness, looseness, lightweight, heaviness, softness, stiffness, anti-drape stiffness, springiness; visual sensations like colour, light, reflection

3. *Socio-psychological factors*: subjective perception, overall perception of sensory sensations from neurophysiologic sensory signals through evaluation of several perceived sensations, previous experiences, aesthetical components like style, texture, aesthetics, fashion, suitability, design and colour, occasion, geographical location, climatic conditions, sociocultural settings and norms, and historical importance, fit, the desire for comfortable clothing and the body image, fashion preferences, cultural and individual preferences

4. *Physiological factors*: oxygen consumption, skin temperature, muscle activity, blood circulation, heart rate, blood pressure, metabolic rate, surface area to volume ratio, age, gender, disability and illness, diet, sleep pattern, activity

5. *Psychophysiological factors*: connectivity of the body and the mind, heart rate variations, brain function, physiological responses

3.2 THERMAL COMFORT: ENVIRONMENTAL, PERSONAL AND CLOTHING PROPERTIES

Study into thermal comfort in humans was born in the early decades of the twentieth century when Adolph Phare Gagge (1908–1993) undertook a series of studies to resolve problems that occurred in stressful workplaces. The period after the war and in the 1970s, with Powl Ole Fanger and other researchers, marked the study of thermal comfort as a credible discipline (Fabbri, 2015).

The first research into convective heat exchange between the human body and environment started in 1930s with Winslow et al.'s study on the partitional calorimetry method, and with Hardy and DuBios's study on the direct calorimetry method based on the assumption of a uniform rate of convective heat exchange across the whole body (Winslow et al., 1936a; Hardy and DuBois, 1938).

In the 1980s, the first calorimeters were produced in order to continuously measure both evaporative heat loss and dry heat exchange from the human body. The respiratory and sweating component of evaporative heat loss was quantified by recirculating air through a system of freezers to absorb water (McLean and Tobin, 1987). Calorimetry is the term applied to the measurement of the exchange of heat. The direct calorimeter measures the heat exchange directly based on a person performing an activity in a controlled environment, while indirect calorimetry involves estimation of the energy exchanges from the measurement of material exchanges. Indirect calorimetry uses the rate of oxygen consumption to estimate metabolic rate. There are two main types of direct calorimeter: one depends on absorbing the body's heat loss and measuring the rise in temperature in the absorbing medium. The other measures the temperature difference produced across a layer surrounding the body as the result of heat flow from the body to its surroundings (Ingram and Mount, 1975; Parsons, 2001). The direct calorimetry is not a practical option for thermal comfort assessment; however, the indirect calorimetry methods also have their flaws, and it is not possible to obtain an accurate estimate of metabolic heat production. The subjects are exposed due to the interference which the equipment has on the activity of the person. There is also a noticeable variation for the same activity between different subjects, variation for the same activity when comparing multiple experiments and problems with calibration and leaks in the equipment. The second issue is that with the high levels of activity the concept of thermal comfort may change due to profuse sweating, blood redistribution and hormonal secretions, which impacts productivity, physical and thermal strain (Parsons, 2001).

The partial calorimetry method was employed much earlier. The basis of partitional calorimetry in application to man was developed by Winslow, Herrington and Gagge. The partial calorimetry method exposes the human subject to an environment in order to separately determine the components responsible for heat exchange. This method allowed the differentiation of each of the heat exchange pathways, convection, radiation, evaporation and conduction, which are very small. The system consists of an enclosure lined with polished copper plates that reflect heat from external radiant heaters onto the subject. Indirect calorimetry methods include the heat sink and the thermal gradient method (one being respiration calorimetry and the other depends on the carbon and nitrogen analysis of food intake) (Winslow et al., 1936a, 1936b, 1937; Ingram and Mount, 1975).

Thermal comfort is a subcategory of overall comfort and regarded as a complex sensation, which integrates various sensory inputs. The sensation of heat, cold and humidity or skin wetness are the main determinants of overall comfort (Havenith, 2002). This is a very multidisciplinary and interdisciplinary field of study (van Hoof et al., 2010). The thermal comfort of a person is defined as a condition of the mind which expresses satisfaction with the thermal environment (Fanger, 1973b) or a state in which there are no driving impulses to correct the environment by changing behaviour (Hensen, 1991).

Thermophysiological wear comfort concerns the heat and moisture transport properties of clothing and the way the clothing helps to maintain the heat balance (Bhatia and Malhotra, 2016).

Within the wide limits of the environmental factors, thermal comfort can be obtained only for a narrow range of those parameters, assuming no clothing or activity changes are allowed (Havenith, 2002). The variables affecting comfort are classified into three major groups: physical variables of the environment and/or the clothing; psychophysiological parameters of the wearer as well as the psychological filters of the brain (Pontrelli, 1977).

The parameters affecting the thermal comfort are classified as (Fourt and Hollies, 1969, 1970; Fanger, 1973a; Gagge and Nishi, 1977; Berglund, 1998; Parsons, 2000; Havenith, 2002; Epstein and Moran, 2006; Kamalha et al., 2013; Angelova, 2016):

1. ***The environmental or external parameters***:
    a. Directly measured parameters (*ASHRAE Handbook of Fundamentals*, Chapter 8 and 14): air temperature (t_a), wet bulb temperature (t_{wb}), dew-point temperature (t_{dp}), water vapor pressure (p_a), total atmospheric pressure (p_t), relative humidity (RH), humidity ratio (W_a) and air velocity (V_a)
    b. Calculated parameters: mean radiant temperature (t_r), radiant plane temperature (t_{pr}), the radiant temperature asymmetry and environmental indices (effective temperature, humid operative temperature, heat stress index, skin wittedness index, wet bulb globe temperature, wet globe temperature, wind chill index)

2. ***The personal factors***:
    a. Human body factors: the metabolic rate of the body, heart rate, body temperature and sweating, DuBois and Dubois surface area, activity level, skin temperature and skin wettedness
    b. Secondary factors (ASHRAE): non-uniformity of the environment and local discomfort, day-to-day variations, seasonal and circadian rhythms, visual stimuli, age, adaptation, sex and outdoor climate
    c. Fabric parameters: thermal resistance, fabric moisture permeability or resistance, water repellency, the speed of drying, air permeability/breathability, mass per unit area, thickness and wind resistance
    d. Clothing parameters: thermal insulation, surface area, weight, bulk and construction/style/design

An increase in the metabolic rate of 17.5 W above resting level is equivalent to a 1-degree increase in the air temperature. The change in the thermal insulation value of clothing of 1 clo is equivalent to a change in the air temperature of 5 degrees at rest and 10 degrees while exercising. However, the change of 1 degree in the mean radiant temperature can be offset by 1 degree in the air temperature, while a 10% change in relative humidity can be offset by a 0.3 degrees in the air temperature. The change of 0.1 m/s in wind speed is equivalent to a change in air temperature of between 0.5 and 1.5 degrees (Epstein and Moran, 2006). Thermal comfort is accomplished through several factors like the wide range of metabolic conditions and environmental and clothing factors.

The assessment of heat stress through a single value combining the several variables is usually invalid since there cannot be a universal valid system for rating heat stress (Gagge and Nishi, 1976), although in recent years the single criterion for acceptability of global thermal comfort has been developed for use in any kind of environment, which may be expressed in terms of the predicted mean vote (PMV) (Lenzuni et al., 2009).

The personal condition for acquiring the body thermal comfort in static conditions have been defined as (Fanger, 1970):

1. The body being in heat balance
2. The sweat rate being within comfort limits
3. The mean skin temperature being within comfort limits

Fanger defined the conditions for static energy balance describing the connection between the human body and his environment in total equilibrium, which implies the resultant of the heat (produced and lost) and the performed work is zero. He also pointed to a strong need for establishing comfort conditions during transients. All of the factors that influence thermal comfort are co-dependent, including the activity level, the clothing and the climate parameters (pertaining derivatives). Derivatives express the change that is produced by a unit change on the variable in the numerator, so that the value of the comfort equation remains zero. Further, Fanger's research moved towards explaining dynamic heat-balance conditions. Through the PMV scale, Fanger introduced the thermal load of the human body, different from equilibrium in the comfort equation. PMV is calculated for different activity levels with various environmental and clothing parameters using the live subjects' votes to classify thermal sensations. Thermal comfort is described as slightly cool, neutral and slightly warm if the votes fall on the scale between −1 and +1. The complete scale has a thermal comfort indicator range between −3 and +3. Following this, further scales have been introduced to explain dynamic heat-balance conditions such as TSENS and DISC scales. Both range from −5 to +5 to describe human tolerance to thermal sensations (Fanger, 1970, 1973a; Fanger et al., 1980; Bánhidi et al., 2008).

Thermal comfort depends on the combined influence of personal and environmental factors. The environmental factors include air temperature, air velocity, mean radiant temperature and air humidity. It is also affected by the heating or cooling of a particular body part due to radiant temperature asymmetry (radiant plane temperature), draft (air temperature, air velocity, turbulence), vertical air temperature differences and floor temperature (surface temperature) (*ASHRAE Handbook of Fundamentals*, 2017). The personal factors which affect thermal comfort include the activity level (internal heat production in the body) and the thermal resistance of clothing. Maintenance of heat balance is far from being a sufficient condition for thermal comfort. (Fanger, 1973a).

Apart from steady-state measurements, modelling of human heat loss is done in a transient, non-uniform environment. Contributing factors for modelling human heat loss in a transient, non-uniform thermal environment include (Farrington et al., 2004):

1. Thermal radiation view factors
2. Thermal radiation, convection and conduction between clothing layers
3. Thermal and moisture capacitance of clothing
4. Clothing to skin contact area
5. Clothing to skin thermal resistance
6. Clothing fit, including micro-volumes
7. Non-uniform thermal properties of clothing ensembles
8. Non-uniform, transient velocity field around the body
9. Modelling the evaporation of sweat

Non-uniform thermal environments are described by two rationally derived indices, the operative temperature and humid operative temperature. The operative temperature is the uniform temperature of an imaginary uniform isothermal black enclosure in which the subject will exchange the same dry heat by radiation and convection as in the actual non-uniform environment. The standard humid operative temperature is the temperature of an imaginary saturated isothermal enclosure in which a human would exchange the same heat by radiation, convection and evaporation as they would in an actual non-uniform environment with the same skin temperature and wettedness, while wearing standard clothing and being exposed to standard air velocity at sea level (Nishi and Gagge, 1971b; Wray, 1980; ANSI/ASHRAE Standard 55-2017).

Heat stress indices are correspondingly categorised into three groups according to their rationale. There are rational indices based on calculations involving the heat-balance equation, empirical indices based on objective and subjective strain, and direct indices based on direct measurements of environmental variables (Epstein and Moran, 2006).

The combined effect of the measured thermal parameters can be calculated by the thermal indices such as the effective temperature determined from air temperature and humidity (Gagge et al., 1971) or the PMV, an index predicting the average thermal sensation that a group of occupants may experience in a given space, based on the all four environmental and personal indices (Fanger, 1982, *ASHRAE Handbook of Fundamentals*, 2017; ISO 7730:2005).

3.2.1 Clothing as a Second Skin: Preserving Thermal Comfort

Comfort is a complex subject and difficult to define. Clothing is in direct contact with the body and interacts continuously and dynamically during wear, stimulating different mechanical, thermal and visual sensations (Li, 2001). Clothing functions as a resistance to heat and moisture transfer between the skin and environment and protects against extreme heat and cold (Havenith, 2002). Clothing comfort provides stimuli to the human body and contributes to the attainment of equilibrium between heat and moisture exchange (Li, 2001; Kamalha et al., 2013).

Clothing can also have major effects on thermal strain, causing health disorders. Any risk assessment procedure explaining the influence of different thermal strains must consider the influence of the thermal properties and effects of the clothing (Havenith et al., 1990).

The diffusion, absorption and desorption, evaporation, wicking, wet conduction (additional conductive heat transfer due to the clothing being wet) and condensation of moisture are all forms of wet heat and moisture transfer in clothing, while conduction, convection, radiation and ventilation are mechanisms of dry heat transfer (Richards et al., 2008).

Dynamic heat and moisture transfer in the whole body–clothing–environment system in transient conditions is somewhat unexplored and falls into the scope of current studies due to the complexity of the dynamic physiological behaviour of human body, the dynamic fabric and clothing properties, clothing ventilation, activity, airflows, postures, etc.

Water from an external source should be prevented from entering the clothing, while the water vapour produced due to metabolic processes should be released from the clothing's microclimate to the outside environment. The water vapour's movement through fabric depends on the type of yarn, the microporous nature of the fabric or in general the structure and application of different finishing treatments (Slater, 1977).

Clothing mediates convective, radiative and evaporative heat exchange between the human body and the environment. The dry heat loss (radiation, convection, conduction) is used to calculate the thermal insulation of the clothing (Nielsen et al., 1985), while the evaporative heat loss is used to calculate the clothing's vapour resistance (Havenith et al., 2008b).

Havenith et al. defined five pathways of heat transmission through the clothing (Havenith, 2005, 2008b):

1. Dry heat pathways due to ventilation
2. Basic dry heat loss
3. Extra conduction in the fabric due to moisture
4. Evaporation of moisture on the skin
5. Evaporation out of the clothing to the environment

Convection heat loss occurs when air cooler than the skin flows along the skin, while heat radiative heat loss occurs if there is a difference between the body's surface temperature and the temperature of the surfaces in the environment. Heat can be lost by

Modelling Heat Losses from the Human Body

conduction in small amounts and by evaporation. There are two types of evaporative processes occurring from the body: the evaporation of moisture (sweat) on the skin or through respiratory evaporative and convective heat losses (Havenith, 2005). Thermal conductivity, thermal resistivity and thermal absorptivity are properties most commonly measured for heat transfer through clothing (Bhatia and Malhotra, 2016).

The rate of heat loss from the human body depends on latent heat loss (evaporative heat loss due to sweating and evaporation) as well as on sensible transfer (non-evaporative heat transfer due to radiation and convection), see Figure 3.2.

The comfort range will move to lower temperatures with the increase in activity level and the increase in clothing insulation (Havenith, 2002). The amount of heat transferred by any pathway is dependent on the specific driving force, the body surface area and the resistance to that heat flow. Conduction, convection and radiation take place at the surface of the body. Conduction is the least pronounced avenue of heat loss. The more important avenues are convective and radiative heat loss. The air around the body allows for convective heat transfer, since the air that flows around the skin is usually cooler than the skin and allows the heat to be transferred to the surrounding air. Radiative heat transfer through electromagnetic radiation can also be substantial if there is a difference between the body's surface temperature and the temperature of the environment surfaces. The last avenue of heat loss is heat loss by evaporation, which is the most prominent way of losing heat in hot environments. Sweat accumulating on the skin's surface can evaporate simultaneously dissipating large amounts of heat from the body. From inside out, the body can also dissipate the heat via respiration. The inhaled air is usually cooler and dryer than the lung's internal surface and after inhaling the lungs warm and moisturise the inspired air thus losing heat with the expired air, which can be up to 10% of the total heat production (Havenith, 2002).

3.2.2 From Fibres to Clothing: Thermal Properties and Applications

Thermal comfort elements related to clothing properties relate to the fabric's aspects, garment design and constitution (Li, 2001; Kamalha et al., 2013).

Thermophysiological comfort is a state of human satisfaction produced by satisfactory heat and moisture transmission through the textile and clothing from the human body to the environment. The fabrics and clothing garments affect the human body inducing the thermophysiological sensations such as coolness, warmth, chilling and sweating. The fibres and material properties influence the overall clothing comfort. When discussing any aspect of comfort, especially clothing comfort, one should consider all of the relevant properties from the types and properties of fibres used, to the types and properties of yarns forming the fabric structure, and all the way to fabric structures and clothing pattern design and layering. Additionally, the dyeing, finishing and coating processes can also influence a material's properties (Dolez et al., 2018).

Fabric properties affecting clothing comfort properties are yarn properties (tenacity, breaking elongation, flexural rigidity), construction parameters, mechanical and surface properties, thermal insulation, air permeability and moisture transfer properties (Behera et al., 1997).

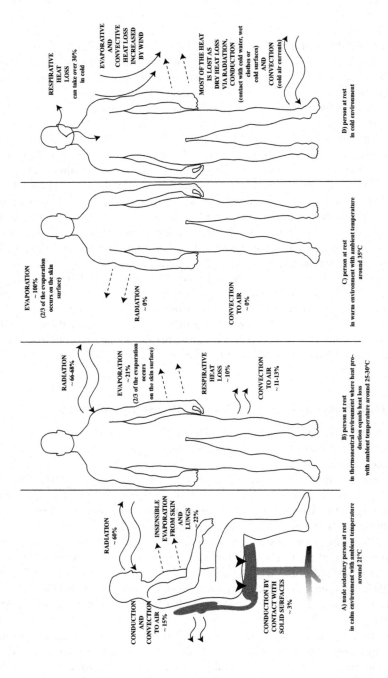

FIGURE 3.2 The heat loss mechanisms in different climatic conditions.

The factors influencing the thermal insulation properties of textiles (Angelova, 2016):

1. Structural factors including the type of fabric (woven, knitted, or non-woven), the characteristics of fabric (density, thickness, mass per unit area, volume mass, porosity, etc.), the types of threads and yarns, the characteristics of threads and yarns (linear density, twist, compact density, etc.), the specific heat of fibres.
2. Surface factors including the texture, hairiness, coatings and films applied to the surface, the contact surface between the macrostructure and the human body.
3. Factors related to dry and latent heat transfer such as heat transfer properties between the fibres and the air, the heat transfer properties between staple fibre yarns or the polyfilaments and the air, the heat loses between the human body and the textile by conduction between the human body and the textile, the heat loses between the skin surface and the surface of the outer textile layer by radiation and, the heat loses between the body and next-to-body textile layer, and between the layers of clothing, by convection, the heat losses by diffusion from the body surface and the outer textile layer surface to the environment, and the heat storage due to absorption or adsorption of moisture from the surface of the textile layer.
4. Environmental parameters (air temperature, mean radiant temperature, operative temperature, surface temperature, radiant temperature asymmetry, relative humidity, air velocity, etc.).

The choice of materials used to produce clothing affects the heat and moisture transfer through the clothing materials, consisting mainly of conduction and radiation. Clothing insulation is thus dependent on the material's properties. The material's thickness and the volume of air enclosed between the fibres and the layers of fabrics affect the insulation and water vapour permeability. Other factors affecting clothing insulation and water vapour involve the type of materials and added substances such as apertures. The fabric's thickness and the thickness of the enclosed air layer overly determine the clothing's vapour resistance in permeable materials. Another major effect of vapour resistance is concerned with the coatings, membranes or other treatments added to the fabrics. The type of fibres used to produce the fabric affects air permeability and moisture absorption, which may affect insulation and vapour resistance in special conditions like high winds and wet environments. Second, since the fibres reflect, absorb and re-emit radiation, they substantially influence the heat transfer via radiation (Slater, 1977; Havenith, 2002).

The fibres providing thermophysiological comfort should ideally have good hygroscopic properties. Good hygroscopic properties are found in some natural fibres such as the wool fibres. Wool and cotton fibres have a higher ability to absorb large amounts of moisture. The fibres are also capable of trapping the still air inside their structure, within the fabric and between the fabric layers thus improving the textile thermal insulation value. The hairy fibres will entrap a greater volume of still air within the fabric structure. A third important property is the size of the fabric pores, which affect the heat and moisture transfer resistance of the fabric and also affect the garment's resistance (Dolez et al., 2018).

The fabrics offer resistance to heat, which is critical for maintaining thermal comfort. However, the fabrics should also be able to provide good air permeability, wind-stopping properties, a combined effect of heat conductivity and moisture diffusivity, moisture and vapour permeability, have absorbent or non-absorbent properties according to specific usage, have water-repellent properties when used as outer layer, provide good liquid-moisture transmission, etc. (Slater, 1977).

Varshney et al. presented a study on fibre effects on the thermophysiological properties of fabrics. They reported that the increase in fibre linear density and use of non-circular fibres cause higher fabric porosity, leading to increased thermal resistance and lower thermal conductivity and absorptivity. A fall in fibre linear density or the use of non-circular fibres increases the spreading speed of a water droplet through the fabric. Coarser and modified profiles of polyester are more comfortable due to the increased in-plane wicking and high porosity, leading to higher water vapour permeability. The pore structure has considerable influence on the fabric's wickability and water vapour permeability. Increased fibre linear density or employing non-circular fibres provides an increase in air permeability (Varshney et al., 2010).

There is a strong correlation between the thermal resistance of the fabrics and the clothing's thermal insulation (Matusiak and Sybilska, 2015). Apart from the choice of materials, the overall clothing garments and the ensembles also affect heat exchange between the body and its environment and the metabolic rate. The thermal resistance of a clothing system represents a quantitative evaluation of the effectiveness of the clothing for providing a thermal barrier to the wearer, while the water vapour permeability of clothing materials is a critical property for a clothing system to maintain the human body at thermal equilibrium (Huang, 2006a).

When considering clothing as a macroscopic formation affecting heat and moisture release from the body and comfort, garment design properties are used to change the transmission rate. These properties are the overall garment fit, weight, ease of movement, pressure to the body's surface, ventilation, elasticity and thermal insulation (Slater, 1977).

In correspondence with the above-mentioned properties, the most important requirements for military and protective clothing were defined as (Scott, 2000):

1. Improved protection against environmental and battlefield threats
2. Improved maintenance of thermophysiological comfort or survival in extreme conditions
3. Improved compatibility and integration between different clothing components
4. Reduction in weight, load carriage and bulk
5. Reduction in life cycle costs, improvements in durability and recyclability

3.2.2.1 Heat Transmission through Textiles and Clothing

Heat and moisture transfer through the garment is not only dependent on the properties of the fabric used in the garment. The properties of textile materials, the

Modelling Heat Losses from the Human Body

properties of air layers formed between the fabrics or different garments (such as the thickness of air layers and the magnitude of the contact area), the properties of the outer adjacent air layer and the properties of the clothing (such as the garment design and the clothing geometry), affect heat and mass transfer in clothing (Psikuta et al., 2016; Mert et al., 2017).

Dry heat transfer at the surface of the skin is the combination of conduction combined with convection and radiation, called sensible heat loss from skin (Song, 2011). However, convection and radiation are the most prominent components of dry heat loss (Havenith, 2002).

Dry heat loss is present with dry skin and correspondingly increases with the temperature gradient between skin and environment. Whereas when the skin temperature rises to 34°C only evaporative heat loss is present and when the skin temperature drops below 34°C evaporative and dry heat loss are present simultaneously (Havenith et al., 2008b).

For static conditions with no air motion, convection and radiation are the two most prominent methods of heat transfer. The combined effect of radiative (R) and convective (C) heat transfer can be written as (Havenith et al., 2002; Holmér et al., 1999; *ASHRAE Handbook of Fundamentals*, 2017):

$$R + C = (t_{sk} - t_{cl})/I_{cl} \quad [m^2K/W] \tag{3.1}$$

or

$$R + C = (t_{sk} - t_o)/I_T \quad [m^2K/W] \tag{3.2}$$

or

$$R + C = F_{cl} \times (h_c + h_r) \times (t_{sk} - t_o) \quad [m^2K/W] \tag{3.3}$$

or

$$R + C = (I_a/I_T) \times (h_c + h_r) \times (t_{sk} - t_o) \quad [m^2K/W] \tag{3.4}$$

or

$$R + C = (t_{cl} - t_o)/(I_a/f_{cl}) \quad [m^2K/W] \tag{3.5}$$

Where t_{sk} is the temperature at the skin's surface (K), t_{cl} is the temperature at the clothing surface (K), t_o is the ambient operative temperature (K), I_{cl} is the clothing intrinsic insulation (m²K/W), I_a is the thermal insulation of the surface air layer (m²K/W), and I_T is the total insulation of clothing and boundary air layer (m²K/W), f_{cl} is the clothing area factor, F_{cl} is clothing efficiency factor, h_c is the convective heat transfer coefficient (W/m²K), and h_r is radiative heat transfer coefficient (W/m²K).

The convective heat transfer (C) can be expressed in terms of the convective heat transfer coefficient (h$_c$) (*ASHRAE Handbook of Fundamentals*, 2017):

$$C = f_{cl} \times h_c \times (t_{cl} - t_a) \quad [W/m^2 K] \tag{3.6}$$

The radiative heat transfer (R) can be expressed in terms of the convective heat transfer coefficient (h$_r$) (*ASHRAE Handbook of Fundamentals*, 2017):

$$R = f_{cl} \times h_r \times (t_{cl} - \overline{t_r}) \quad [W/m^2 K] \tag{3.7}$$

The steady-state heat transfer through fabric is governed by Fourier's law of thermal conduction (Huang, 2016):

$$K = (k \times A \, \Delta T)/L \quad [W] \tag{3.8}$$

Where K is the thermal conductive heat flow rate (W), k is the thermal conductivity (W/mK), A is area (m²), ΔT is the temperature gradient (K), and L is the thickness of the fabric (m).

It can also be written as (*ASHRAE Handbook of Fundamentals*, 2017):

$$K = -(k \times A) \times (dt/dx) \quad [W] \tag{3.9}$$

or

$$K = k \times (A/L_m) \times \Delta T \quad [W] \tag{3.10}$$

Where K is the thermal conductive heat flow rate (W), k is the thermal conductivity (W/mK), A is the cross-sectional area normal to flow (m²), (dt/dx) is the temperature gradient (K/m), and L$_m$ is the mean length of heat flow path (m).

From the above-mentioned equations, the thermal resistance can be derived from (*ASHRAE Handbook of Fundamentals*, 2017; Huang, 2016):

$$K = dT/R, \text{ from which } R = dT/K = 1/k \times L/A \quad [K/W] \tag{3.11}$$

Thermal resistance is the temperature difference between the two faces of a material divided by the resultant heat flux per unit area in the direction of the gradient (BS EN ISO 14058, 2017). Thermal resistance R is directly proportional to the mean length of the heat flow path (which can be seen as the thickness of the fabric) and inversely proportional to conductivity k and the mean cross-sectional area A normal to the flow.

The heat resistance (R$_{ct}$) of fabrics is generally related to the thermal conductivity of the fibres and fabric thicknesses (L), but is also dependent on the relative porosity (Bedek et al., 2011):

$$R_{ct} = h/K \quad [m^2 K/W] \tag{3.12}$$

Modelling Heat Losses from the Human Body

Thermal resistance of the textile materials or composites (R_{ct}) can be written as the quantity, which determines the dry heat flux across a given area in response to a steadily applied temperature gradient (BS EN ISO 14058, 2017; ISO 11092:2014):

$$R_{ct} = \left\{\left[(T_m - T_a) \times A\right]/H\right\} - R_{ct0} \quad \left[m^2 K/W\right] \quad (3.13)$$

Where (T_m) is the temperature of the measuring unit at 35°C and (T_a) is the air temperature at 20°C, H is the heating power supply to measuring unit, A is the area of the measuring unit, R_{ct0} is heat resistance measured without a sample present, measured with sweating guarded-hotplate apparatus under a relative humidity of 65% and the airspeed v_a is 1 m/s.

The resistance of the fabric is lower than the convective resistance of the air it displaces (Lotens and Havenith, 1989). The overall thermal resistance of clothing is affected by the fabric material type, thickness, relative humidity and air or wind velocity. The fabrics made from wool fibres, acrylics, piles and special fabrics made from textured yarns have greater insulating properties. Controlled ventilation and layering in clothing provides an increase in the overall clothing thermal resistance (Li, 2001; Kamalha et al., 2013). The fabric's thermal resistance represents a quantitative evaluation of the thermal barrier provided by the fabric, while the water vapour resistance of the fabric is a critical property since the high water vapour permeability in cold environments can minimise water accumulation in clothing, which leads to an increasing sense of discomfort. In general, the thermal resistance of fabrics and the thermal insulation of clothing is merely the resistance offered to the flow of heat. It is measured by the ratio of the difference in temperature between two surfaces, to the flow of heat per unit area (Gagge and Burton, 1941; Huang, 2006b).

The energy and insulation values can be expressed as met and clo units. One clo unit represents the value of the everyday clothing insulation, while one met represents an energy consumption for a man of average size and is approximately equivalent to the heat generated by a 100-W lamp (Gagge et al., 1941).

One clo of insulation will balance the heat produced by a resting man under standard indoor climatic conditions (Burton and Edholm, 1955; Holmér et al., 1999):

$$1 \, clo = 0.1555 \, m^2 K/W. \quad (3.14)$$

The total insulation (I_T) is defined as thermal insulation from the body surface to the environment (including all clothing, enclosed air layers and the boundary air layer) under reference static conditions (ISO 9920:2008; ISO 15831:2004):

$$I_T = (\bar{t}_{sk} - t_o)/H \quad \left[m^2 K/W\right] \quad (3.15)$$

or

$$I_T = I_{cl} + (I_a/f_{cl}) \quad \left[m^2 K/W\right] \quad (3.16)$$

Where $\overline{t}_{sk}$ is the mean skin surface temperature (K), t_o is the operative temperature (K), f_{cl} is the clothing area factor calculated as the ratio of the outer surface area of the clothed body to the surface area of the nude body (-), H is the dry heat loss per square metre of skin (W/m²), I_{cl} is the basic insulation or intrinsic clothing insulation and is defined as the thermal insulation from the skin surface to the outer clothing surface (including enclosed air layers) under reference static conditions (m² K/W) and I_a is thermal insulation of the boundary (surface) air layer around the outer clothing or the unclothed body (m² K/W).

The standard operative temperature is the temperature of a black isothermal environment in which a human being, wearing standard clothing and exposed to a standard air movement and barometric pressure, would exchange the same heat by radiation and convection as it would with the same mean skin temperature in the actual environment worn (Gagge and Nishi, 1977). The operative temperature is one of the rationally derived indices for characterising non-uniform thermal environments, along with the humid operative temperature (Wray, 1980).

The basic insulation or intrinsic clothing insulation (I_{cl}) is defined as the thermal insulation from the skin surface to the outer clothing surface (including enclosed air layers) under reference static conditions (ISO 9920:2008; ISO 15831:2004):

$$I_{cl} = (\overline{t}_{sk} - t_{cl})/H = I_T - (I_a/f_{cl}) \quad [m^2 K/W] \qquad (3.17)$$

Where t_{cl} is the clothing surface temperature (K). The expression (I_a/f_{cl}) accounts for the insulative effect of the body when the air insulation is reduced, the thicker the clothing or, better to say, the larger the outer clothing surface area.

Air insulation (I_a) is defined as thermal insulation of the boundary (surface) air layer around the outer clothing or the unclothed body (ISO 9920:2008; ISO 15831:2004):

$$I_a = (\overline{t}_{cl} - t_o)/H = 1/(h_c - h_r) \quad [m^2 K/W] \qquad (3.18)$$

Where h_c is the convective heat transfer coefficient (W/m²K), and h_r is the radiative heat transfer coefficient (W/m²K).

The resultant or dynamic total insulation ($I_{T,r}$) is defined as actual thermal insulation from the body surface to the environment (including all clothing, enclosed air layers and boundary air layers) under given environmental conditions and activities. It corresponds to the total insulation but is measured with the thermal manikin moving and thus accounts for the effects of movement and wind on the insulation reduction (Holmér, 1989; ISO 9920:2008; ISO 15831:2004).

The resultant basic clothing insulation ($I_{cl,r}$) is defined as the actual thermal insulation from the body surface to the outer clothing surface (including enclosed air layers) under given environmental conditions and activities. It corresponds to basic clothing insulation but is measured with the thermal manikin moving and thus accounts for the effects of movement and wing on the insulation reduction (ISO 9920:2008; ISO 15831:2004).

Modelling Heat Losses from the Human Body

The effective thermal insulation of a clothing garment (I_{clu}) is the increase in insulation provided to a thermal manikin by a single garment compared to the nude manikin's insulation (ISO 9920:2008; BS EN ISO 15831:2004):

$$I_{clu} = I_T - I_a = \left[(\overline{t}_{sk} - t_o)/H\right] - I_a \quad \left[m^2 K/W\right] \quad (3.19)$$

Since the garment is defined as the individual component of a clothing ensemble, which provides protection to the part of the body that it covers, the clothing ensemble is defined as the number of garments covering the body, except head, hands and feet. It can also be one garment covering the entire body such as coverall (ISO 14058:2017).

The effective thermal insulation of clothing ensemble (I_{cle}) is the increase in insulation provided to a thermal manikin by a whole ensemble (the combination of garments worn simultaneously) compared to the nude manikin insulation (ISO 9920:2008; ISO 15831:2004):

$$I_{cle} = I_T - I_a \quad \left[m^2 K/W\right] \quad (3.20)$$

I_{cle}, however, neglects the effect of the greater surface area on the outside of the clothing (ISO 9920:2008).

Heat transfer from the human body depends on the amount of ventilation between the skin and the clothing layers and the microclimate volume (fit) and both the combination of the body movement, posture and wind (Havenith et al., 1990). One should consider the type of garments, the number of layers in the clothing ensemble and the properties of the air layers formed between garments (Havenith, 2002). Human subjects tend to modify their clothing to alter their heat exchange and thermal condition up to several times a day, even in the indoor environments (Morgan and de Dear, 2003).

The geometric factors also have a substantial impact on the overall clothing insulation since clothing is not a single homogeneous layer that covers the body as previously considered. The number of layers over different segments of the body vary, and the insulation is not uniformly distributed over the body. The outermost layers have less effective insulation due to their larger surface area. The ensembles entrap more air than fabric alone and provide greater ability for the radiative heat transfer, in particular for loosely fitting clothing (Lotens, 1989).

The still layer of air formed around the surface of the clothing is around 6 mm thick. The insulation or vapour resistance can be expressed in terms of an equivalent to the still air thickness. The resistance for heat or vapour transport over the body is calculated as the sum of the thickness of the still air layer trapped between skin and clothing, the still air equivalent of material and the still air layer on the outside of the clothing. The clothing ensemble consisting of several material layers will produce an insulation equivalent higher than the insulation of a single material layer; however, air layer thickness is affected by the clothing's fit, body posture, movements and the wind around the body (Belding et al., 1947).

Bending due to different body postures compresses the garment, while flexing elongates the garments (Kirk and Ibrahim, 1996). Posture affects body dimensions and the garment's geometry. Clothing can restrict body mobility and change body shape via a compression. In the last few years, the 3D body scanning technology and reverse engineering techniques allowed the extensive analysis of human body geometry and the changes due to different body postures (Petrak et al., 2015; Mert et al., 2016; Mahnic Naglic and Petrak, 2017).

The insulation values of clothing are affected by the fit since the unventilated trapped air layer is a major contributor to clothing insulation. The ventilation of the trapped air and the air movement (pumping effect) generated within the garment due to either activity or outside air motion also affects the clothing insulation due to higher convective heat transfer through the clothing (Rohles et al., 1973). Loose-fitting ensembles entrap more air than tight-fitting garments, which helps to encourage a larger reduction of the clothing's insulation value due to the movement, posture and wind (Havenith et al., 1990).

The heat and vapour transfer properties of the clothing are highly important. The second major significance is the ventilation properties of the clothing and the effect of the workers' activity and environment on the thermal properties (Havenith et al., 1990). Zhang and Wang released a study on the importance of clothing ventilation openings. The results point to a greater role of neck openings on heat transfer and ventilation during exercise than hem openings. This has proven even more important during the recovery period after exercise (Zhang and Wang, 2012).

The inhomogeneous thickness of the air layers within the clothing system influence the local heat and vapour exchange. The EMPA research team did an evaluation of the air gap thickness and contact area between the body and garment using 3D body scanning technology and scanned post-process using Geomagic Qualify 11 software. The results suggested that regardless of garment fit, the clothing provided smaller air gaps and a larger contact area on the upper torso but strongly influenced air layer thickness and contact area on the lower torso (Frackiewicz-Kaczmarek et al., 2015; Psikuta et al., 2015). Mert et al. proved that the thickness of the air layer decreased below the threshold of natural convection and the magnitude of the contact area increased in many postures on both upper and lower body garments. Posture changed the distribution of air gap thickness and the contact area. However, heat will be transferred mainly by conduction and radiation over the entire body for regular fitting garments and on the upper trunk for loose-fitting garments regardless of the selected body posture, due to an air gap thickness below the threshold of natural convection (Mert et al., 2017). This could be the consequence of the somewhat limited selection of upper body postures, differing only in an elbow flexion between 90 to 120°, and maximum shoulder flexion of 30°. The selected postures of the upper body had an effect on the air gap thickness only at the abdomen, lumbus and pelvis area.

3.2.2.2 Moisture Transmission through Textiles and Clothing

Moisture transfer in general plays an important role in forming human perception of overall comfort and especially on the thermal comfort. The accumulation of sweat on the skin adds to a thermal discomfort sensation (Winslow et al., 1939) as well as the poor evaporation of the sweat through the clothing (Takada et al., 2004).

The skin wettedness is the fraction of the skin covered with water to account for the observed total evaporation rate (Gagge, 1937). The water vapour that accumulates and absorbs into the fabrics influences the thermo-physical properties such as thermal conductivity, heat capacity, evaporative heat loss and also generates the phase changed energy. The influence of moisture on thermal protection is the result of a balance between the positive effects of water due to the higher heat capacity and energy absorption through vaporising and the negative effects due to the higher heat conductivity of the wet material (Wang et al., 2012). If the clothing is unable to provide an optimum rate of sweat evaporation from the skin, condensation of the sweat will occur. Heat transfer changes during condensation or absorption from that of dry and wet heat transfer (Lotens and Havenith, 1992).

The textile fabrics are considered as porous materials. There are the two main processes happening simultaneously during the thermal drying of a wet textile material. One is the transfer of heat to raise wet porous media temperature and to evaporate the moisture content. Simultaneously, the transfer of mass in the form of internal moisture to the surface of the porous material provides the means for subsequent evaporation (Haghi, 2011).

The modes of vapour and moisture transport in porous materials, among other textiles, are (Bejan, 2004):

1. Transport by liquid diffusion
2. Transport by vapour diffusion
3. Transport by effusion (Knudsen-type diffusion)
4. Transport by thermos-diffusion
5. Transport by capillary forces, which is the most prominent in textile materials
6. Transport by osmotic pressure
7. Transport due to the pressure gradient

Wet transfer mechanisms in textile materials include two major processes, vapour diffusion in the void space and moisture sorption by the fibre, evaporation, and capillary effects. Because of water vapour concentration differences, the water vapour will diffuse and move through the textile fibres. There is fibre and moisture molecular attraction at the surface of the fibre materials. Surface tension and effective capillary pore distribution and pathways affect this molecular attraction. The liquid–moisture transmission is the result of the absorption in the fibres due to their internal chemical compositions and fibre and molecular moisture attraction at the surface of fibre materials. The heat transfer process in textile material is coupled with the moisture transfer processes with phase changes such as moisture sorption/desorption and evaporation/condensation. Thus, depending on the temperature and moisture distributions, the evaporation and/or condensation in textile material can take place (Haghi, 2011).

If the moisture from clothing is transferred in vapour form, a mechanism like diffusion, sorption, absorption, moisture convection and condensation are involved. If the moisture from clothing is transferred in liquid form, a mechanism like wetting and wicking take place. Diffusion is the transfer of water vapour molecules due to kinetic energy from random movement, which occurs along the fibres and through the air spaces between the fibres and yarns. Sorption–desorption happens during transient conditions when the water bound to the textile fibres releases again as vapour carrying the heat of swelling plus the heat of evaporation, simultaneously reducing the local temperature. Absorption–adsorption will take place with the water vapour travelling through textiles. The materials absorb vapour until an equilibrium is reached. The heat released by absorption is composed of the heat of condensation and the heat of swelling, raising the local temperature. The moisture convection is similar to dry heat loss by convection when the ventilation occurs. The moving air takes the microclimatic moisture, which is then replaced by fresh air in the convective stream. Condensation usually occurs with low atmospheric temperatures, as a direct result of the local vapour pressure rising to saturation vapour pressure at the local temperature in the saturated fabrics. The fabrics acting as a cold wall will enhance condensation of the warm and moist air from the body. Wetting is a process of the fluid spreading when the fibre–air interface is replaced with a fibre–liquid interface due to the Young–Dupre law. Wicking is the process of moisture (sweat) spreading from the skin through the fabric offering a dry feeling and enhanced evaporation. When the fibre is wet, the liquid reaches the capillary, and a pressure is developed which forces the liquid to wick or move along the capillary. The liquid reaching the spaces between the fibres increases the capillary pressure according to the Laplace law, and the liquid flows under capillary pressure in horizontal capillaries, trawling the distance calculated by the Washburn–Lukas equation. The different forms of wicking are transplanar or transverse wicking, in-plane wicking and vertical or longitudinal wicking (Bhatia and Malhotra, 2016).

Condensation will appear if the vapour concentration is increased to the point of saturation in the clothing system. In cold conditions, the condensation generally takes place where the temperature gradient or the permeability of the clothing drops (Yoo and Kim, 2012). The water vapour transfer is slow if resistance to water vapour diffusion is high, and the discomfort sensation of dampness and clamminess may arise (Zhang et al., 2001; Holmér, 2006). Moisture is usually ventilated through the clothing's microclimate and diffused through the outer clothing layer. However, the moisture can condense in outer clothing layers if their temperature is lower than the skin temperature. The higher the fabric moisture content, the higher is the heat loss due to the increase of the wet thermal conductivity and the water evaporation (Havenith et al., 2008b).

The wet heat loss is also called latent or evaporative heat loss. The evaporative heat loss calculation is essential to heat-balance calculations; however, the value for the latent heat of evaporation may not always reflect the real cooling benefit to the body. The sweat evaporation is considered to be the most important mode of heat loss during environmental temperatures rising above the standard conditions or when heat loss is limited in protective clothing (Havenith et al., 2013).

Evaporation of sweat is the most important avenue of body heat loss during exposures to hot environments or during intense activity. The fabrics must allow water vapour to escape in time to maintain the relative humidity between the skin and the first layer of clothing at about 50% (Zhang, et al., 2001; Holmér, 2006). The moisture evaporating from the skin can either diffuse through the clothing, pass through

Modelling Heat Losses from the Human Body

the clothing openings in vapour form or condense within the clothing (Richards et al., 2008).

Evaporation is the process of liquid water passing directly to the vapour phase. In the case of the human body, this corresponds to the sweat evaporating from the skin to the environment. The factors influencing the evaporation process in on nude body include the energy supply for vaporisation, called the latent heat, and the transport of vapour away from the evaporative surface, which includes the wind velocity over the surface and the specific humidity gradient above the surface. In the case of the clothed human body, this air velocity is formed in the clothing's microclimate and the gradient between the skin water vapour pressure (p_{sk}) and the ambient air-water-vapour pressure (p_a). The factors influencing the evaporative heat transfer between the body and the environment can be divided into body factors, which are surface or skin temperature, the percentage of the wetted area and site of evaporation relative to the skin surface; and the environmental factors, which are humidity, air velocity and direction. When the net sensible (non-evaporative) heat exchange becomes zero, the only means of dissipating additional heat is by evaporative heat loss (Ch.2, Ingram and Mount, 1975).

An increased evaporative heat loss dominates with permeable clothing; however, the wet material absorbs more radiative heat, especially when it is hydrophobic, and processes transferring heat to the body by the re-condensation of moisture at inner layers of the clothing become apparent with impermeable garments (Bröde et al., 2009).

Evaporative heat transfer by vaporisation in relation to the vapour pressure difference between the skin and the ambient is the driving force, similar to the temperature difference between the skin and the ambient in sensible heat transfer. Water vapour pressure is a measure of absolute humidity (also called the water vapour density, which consists of the mass of water vapour per unit volume of air), while the relative humidity is a measure of the proportion of the saturation vapour pressure represented by the water vapour already present (the ratio of the partial pressure of water vapour in the mixture to the saturated vapour pressure of water at the given temperature). When the water vapour pressure remains constant, the relative humidity decreases as the temperature rises, because of the higher saturation of vapour pressure (Ingram and Mount, 1975).

The total evaporation also called the wet or moisture thermal resistance of the clothing depends on different factors, and on the thickness of the fabrics (Wilson et al., 2000). The total wet heat loss is the combination of three major pathways. It combines the extra conduction in the fabrics due to moisture, the evaporation of moisture at the skin that re-condensates in clothing and releases heat to the outer clothing layers (called the microclimate heat pipe), and one real evaporative pathway, which accounts for the evaporation out of the clothing to the environment. The heat pipe heat loss in the microclimate results from combined conductive heat loss and the evaporation–condensation process in the microclimate (Havenith, 2009).

The total evaporative heat loss E is the combination of the evaporative heat loss of sweat from the skin (E_{sk}) and respiratory evaporative heat loss (Er_{es}). E_{sk} is the total rate of evaporative heat loss through the sweat from the skin (W/m^2). The latent heat loss can be determined by both direct calorimetry (Holmér and Elnäs, 1981) and partitional calorimetry (Gagge and Gonzalez, 1996).

The steady-state water vapour transfer through the fabric is defined by Fick's law of mass diffusion through solids (Gibson et al., 1995; *ASHRAE Handbook of Fundamentals*, 2017) where the mass flow with time (Dincer and Rosen, 2013; *ASHRAE Handbook of Fundamentals*, 2017):

$$\dot{m} = dm/dt = \rho \times A \times v_e \quad [g/s]. \tag{3.21}$$

Where $\dot{m} = dm/dt$ is the mass change per unit of time (rate of mass loss) and equals the mass flow with time (g/s), ρ is the saturation water vapour density (g/m^3), A is the surface area (m^2), v_e is the velocity of evaporation of the water vapour (m/s).

The heat flux is also called the thermal flux, heat flux density, heat transfer per unit area, heat flow per unit area or heat rate per unit area (Balmer, 2011; Dincer and Rosen, 2013; *ASHRAE Handbook of Fundamentals*, 2017). Fick's law can be simplified for cases of dilute mass diffusion in solids, stagnant liquids, or stagnant gases (*ASHRAE Handbook of Fundamentals*, 2017):

$$\dot{m}_{WV} = -D_v \times (d\rho_{wv}/dy) \quad [g/s]. \tag{3.22}$$

Where $\dot{m}_{WV}$ is the total mass flux (g/m^{2}s), also called the mass diffusion flux of water vapour (g/s), $d\varphi_{wv}/dy$ is the gradient of water vapour in the void space or the partial pressure gradient formulation for mass transfer is in direction y (g/m^2), and D_v is the mass diffusivity or the diffusion coefficient (m^2/s).

In this case, the water vapour density is smaller than the density of moist air (air and water vapour mixture), and velocity is almost negligible. This equation can be used for water vapour diffusing through the air at atmospheric pressure and a temperature less than 300.15 K (27°C) if the water vapour density is lower than 0.02 g/m2. This shows that mass flow with time depends on density and wind speed. Fick's law can be written in terms of a partial pressure gradient instead of a concentration gradient for cases of dilute mass diffusion. Diffusion of water vapour is due to the concentration gradient; however, when the water vapour diffuses from the liquid surface into surrounding air, it is assumed that local equilibrium exists through the gas mixture, that the gases are ideal, and that the Gibbs–Dalton law is valid, which implies that the temperature gradient has a negligible effect. Diffusion of water vapour is thus only due to the continuous gas phase, so the mixture pressure (p) is constant (*ASHRAE Handbook of Fundamentals*, 2017).

Fick's law of diffusion of water vapour through the fabric can be written as (Huang, 2016; Gibson et al., 1995; *ASHRAE Handbook of Fundamentals*, 2017):

$$\dot{m}/A = \overline{\Delta C}/R_{df}, \tag{3.23}$$

Where $\dot{m}$ is the mass flux of water across the sample (kg/s), A is the area of the test sample (m^2), $\overline{\Delta C}$ the water vapour concentration difference (kg/m^3) and R_{df} is the water vapour diffusion resistance (s/m).

Modelling Heat Losses from the Human Body

Mass transfer in the drying of wet porous materials, such as textiles, will depend on two mechanisms. These are the movement of moisture within the porous material and the movement of water vapour from the material surface as a result of the external conditions of temperature, air humidity and flow, area of the exposed surface and supernatant pressure (Haghi, 2011).

Mass transfer can happen by either molecular diffusion or convection, which is the transport of one component of a mixture relative to the motion of the mixture and is the result of a concentration gradient (*ASHRAE Handbook of Fundamentals*, 2017). The mass flux or flow rate per unit area of water vapour diffusing through the fabric from one side of the to the other can be calculated as (Gibson et al., 1995; *ASHRAE Handbook of Fundamentals*, 2017):

$$G = \dot{m}/A = (\dot{V} \times \Delta C)/A = (dm/dt)/A = \rho \times v_e \quad [g/m^2 s]. \tag{3.24}$$

Where $\dot{m}$ is the mass flux of water vapour across the sample (g/s), ρ is the saturation density for water (g/m³), A is the surface area of the fabric (m²), $\dot{V}$ is the volumetric flow rate (m³/s) and ΔC is the water vapour concentration difference (g/m³).

The evaporative heat exchange is also affected by convection as explained by the Lewis relation, however, only for the uncovered surfaces and it fails to describe the complex heat exchange in all of the clothing (Holmér et al., 1999).

The evaporative skin loss via skin diffusion (E_{dif}) is calculated as (*ASHRAE Handbook of Fundamentals*, 2017; Fanger, 1970; Atkins and Thompson, 2000; Havenith et al., 2008b):

$$E_{dif} = \left[\lambda \times m \times A_D \times (p_{sk} - p_a)\right]/A_D = \lambda \times m \times (p_{sk} - p_a) \tag{3.25}$$

or

$$E_{dif} = (1 - w_{sw}) \times 0{,}06 E_{max}. \tag{3.26}$$

Where p_{sk} is the saturated water vapour pressure at the skin surface (kPa), p_a is the water vapour pressure in the air flowing over the clothing (kPa), m is water vapour permeance coefficient of the skin (ng/s·m² Pa), also called the permeation coefficient. The water vapour permeance is the rate of water vapour transmission by diffusion per unit area of a body between two specified parallel surfaces, induced by unit vapour pressure difference between the two surfaces (*ASHRAE Handbook of Fundamentals*, 2017), λ is heat of vaporisation of water or enthalpy of evaporation, 2.43 J/g at 30°C, $w_{sw} = E_{sw} - E_{max}$ is the portion of a body that must be wetted to evaporate the regulatory sweat, and the term $(1-w_{sw})$ signifies the portion of skin not covered with sweat (*ASHRAE Handbook of Fundamentals*, 2017), w is the skin wettedness.

When there is no regulatory sweating, the skin wettedness due to diffusion is approximately 0.06 for normal conditions. With regulatory sweating, the 0.06 value applies only to the portion of the skin not covered with sweat $(1-w_{sw})$ (*ASHRAE Handbook of Fundamentals*, 2017).

The maximum possible evaporative heat loss or maximum evaporative potential (E_{max}) occurs with completely wet skin (w = 1).

$$E_{max} = [w(p_{sk} - p_a)]/\{R_{e,cl} + [1/(f_{cl} \times h_e)]\}$$
$$= (p_{sk} - p_a)/\{R_{e,cl} + [1/(f_{cl} \times h_e)]\}. \quad (3.27)$$

The evaporative heat loss by regulatory sweating is directly proportional to the regulatory sweat output (Prek, 2006). The evaporative skin loss via the evaporation of sweat is secreted due to thermoregulatory control mechanisms, also called regulatory sweating (E_{sw}), and is calculated as (*ASHRAE Handbook of Fundamentals*, 2017; Atkins and Thompson, 2000; Havenith et al., 2008b):

$$E_{sw} = \dot{m}_{sw} \times \lambda \quad (3.28)$$

or

$$E_{sw} = w_{sw} \times E_{max} \quad (3.29)$$

or

$$E_{sw} = \langle\{(\Delta m_g - (\Delta m_{wat} + \Delta m_{sol})) - (0.019 \times V_{O_2} \times (44 - p_a)) \times t\} \times 2430\rangle / [(t \times 60) A_D] \quad (3.30)$$

Where Δm_g is the gross body mass loss (g), Δm_{wat} is the mass variation of the body due to intake and excretion of water (g), Δm_{sol} is the mass variation of the body due to intake and excretions of solids (g), the term $0.019 \times V_{O_2} \times (44-p_a)$ is the respiratory weight loss in (g/min), where V_{O_2} is oxygen uptake (L/min), t is observation time (min), and A_D is a body surface area (m²), $\dot{m}_{sw}$ is rate at which regulatory sweat is generated (kg/s·m²) and λ is heat of vaporisation of water or enthalpy of evaporation, 2.43 J/g at 30°C.

The total evaporative heat loss from skin (E_{sk}) can be calculated as the sum of the evaporative skin loss due to sweat evaporation from the skin surface (E_{sw}) and evaporative skin loss due to skin diffusion (E_{dif}) (Holmér, 1995; *ASHRAE Handbook of Fundamentals*, 2017; Havenith et al., 2008b; ISO 9920:2008):

$$E_{sk} = E_{sw} + E_{dif} \quad (3.31)$$

or

$$E_{sk} = [\times(p_{sk} - p_a)]/\{R_{e,cl} + [1/(f_{cl} \times h_e)]\} \quad (3.32)$$

or

$$E_{sk} = [w \times (p_{sk} - p_a)/R_{e,T}] \quad (3.33)$$

Modelling Heat Losses from the Human Body

or

$$E_{sk} = w \times F_{pcl} \times f_{cl} \times h_e \times (p_{sk} - p_a) \quad (3.34)$$

or

$$E_{sk} = w \times h_e \times (p_{sk} - p_a) \quad (3.35)$$

or

$$E_{sk} = (\lambda/A_D)/(dm/dt) \quad (3.36)$$

Where p_{sk} is the water vapour pressure at the skin surface (kPa), p_a is the water vapour pressure in the air flowing over the clothing (kPa), $R_{e,T}$ is the total evaporative resistance of the clothing ensemble and surface air layer, h_e is the evaporative heat transfer coefficient (W/kPa m²), F_{pcl} is permeation efficiency (Nishi and Gagge, 1971a), the ratio of the actual evaporative heat loss to that of a nude body in the same conditions, including an adjustment for the increase in surface area due to the clothing. $R_{e,cl}$ is the basic evaporative heat transfer resistance of the clothing (W/kPa m²), or to be exact from the body surface to the outer clothing surface, f_{cl} is clothing surface area factor (–), w is the skin wettedness factor (–), $\dot{m} = dm/dt$ is the body mass change per unit time and equals the mass flow with time (g/s), λ is the heat of vaporisation of water or enthalpy of evaporation, 2.43 J/g at 30°C, also called the latent heat of sweat evaporation at the skin temperature (J/g) and A_D is the body surface area (m²).

The evaporative resistance is also called the resistance to evaporative heat flow and can be calculated as a quantity that determines the latent or evaporative heat flux of a textile layer under steady-state conditions affected by a partial water vapour pressure gradient perpendicular to the fabric (Umbach, 1988). $R_{e,T}$ is the total evaporative resistance of the clothing ensemble and surface air layer (including all clothing, enclosed air layers and boundary air layers) or the total equivalent uniform impedance to the transport of water vapour from the skin to the environment (kPa m²/W). Instead of measuring the heat flux and mass, loss due to sweating is performed, no regional resistances can be calculated other than the total evaporative resistance, which equals (ISO 9920:2008):

$$R_{e,T} = R_{e,cl} + [1/(f_{cl} \times h_e)] \quad [\text{kPa m}^2/\text{W}] \quad (3.37)$$

or

$$R_{e,T} = [(p_{sk} - p_a) \times A]/H_e \quad [\text{kPa m}^2/\text{W}] \quad (3.38)$$

or

$$R_{e,T} = [(p_{sk} - p_a) \times A]/(\Delta m_e \times \lambda) \quad [\text{kPa m}^2/\text{W}] \quad (3.39)$$

Where H_e is the heat loss from the skin (W), p_{sk} is the saturated water vapour pressure at the manikin's sweating surface (kPa), p_a is the water vapour pressure in the air flowing over the clothing (kPa), A is the area of the manikin's surface that is sweating (m²), λ is the heat of vaporisation of water at the measured surface temperature (J/g), and Δm_e = dm/dt is the mass loss due to the evaporated sweat or the amount of moisture evaporated from the ensemble per unit of surface area per second (g/m²s) or simply the evaporation rate of moisture leaving the manikin's sweating surface and shall not include any water that drips from the surface of the manikin or from the clothing or water absorbed by the clothing.

The clothing's total evaporative resistance is the limiting factor for sweat evaporation from the skin, since only clothing with low evaporative resistance provides the ability of sweat evaporation. Clothing with a low permeability and high evaporative resistance will prevent the evaporated sweat from flowing through the clothing. Permeability is the ability of water vapour to move through clothing affecting the amount of evaporative cooling. If the clothing is impermeable or the sweat secretion is too severe, the sweat will start to accumulate and absorb into the clothing, adding to the sensation of dampness (Li et al., 1995; Plante et al., 1995). Clothing, according to this attribute, is divided into three major groups, permeable, semi-permeable and impermeable.

Havenith et al. proved that total manikin heat loss is affected by the temperature gradient between the skin and the environment and by the effect of the outer garment vapour permeability index and vapour resistance ($R_{e,cl}$) (Havenith et al., 2008b).

The water vapour resistance of clothing ($R_{e,cl}$), also called the basic evaporative heat transfer resistance of the clothing, can be measured through heat-balance analysis in a wear trial with human subjects or with a sweating thermal manikin (ISO 11092:2014):

$$R_{e,cl} = \left\{\left[(p_m - p_a)\times A\right]/H\right\} - R_{e,0} \quad \left[\text{kPa}\,\text{m}^2/\text{W}\right] \quad (3.40)$$

Where p_a is the water vapour partial pressure of the air (Pa) in the test enclosure at the air temperature t_a of 20°C, p_m is the saturation water vapour partial pressure at the surface (Pa) of the measuring unit at temperature t_m of 308.15 K (35°C), H is the corrected heating power supply to measuring unit, A is the area of the measuring unit (m²), $R_{e,0}$ is vapour resistance (kPa m²/W) measured without a sample present (bare plate resistance), measured with sweating guarded-hotplate apparatus under a relative humidity of 65% and the air velocity v_a is 1 m/s.

$R_{e,cl}$ is the basic evaporative heat transfer resistance of the clothing, from the body surface to the outer clothing surface (including enclosed air layers), the impedance to transport of water vapour of a uniform layer of insulation covering the entire body that has the same effect on evaporative heat flow as the actual clothing (kPa m²/W), and equals (ISO 9920:2008):

$$R_{e,cl} = R_{e,T} - \left(R_{e,a}/f_{cl}\right) \quad \left[\text{kPa}\,\text{m}^2/W\right] \quad (3.41)$$

Where $R_{e,T}$ is the resultant total water vapour resistance (kPa m²/W), $R_{e,a}$ is the air-water vapour resistance (the vapour resistance of the surface air layer above the surface of the textile material (kPa m²/W) and f_{cl} is the clothing surface area factor.

Modelling Heat Losses from the Human Body

The clothing surface area factor f_{cl} is defined as the ratio of the surface area of a clothed manikin or person to the surface area of the nude manikin or person (A_{cl}/A_D). It can be predicted by equations (McCullough and Jones, 1984).

Havenith et al. proposed calculation for the evaporative cooling potential (E_{mass}), which is the calculated latent heat content of the moisture that is evaporating from the human–clothing system as measured by the mass loss rate per unit area (Havenith et al., 2008b).

The evaporative cooling potential (E_{mass}), equals the evaporative heat loss from the skin (*ASHRAE Handbook of Fundamentals*, 2017), and can be calculated as the simple heat flux (rate of heat flow):

$$E_{mass} = (\lambda/A_D)/(dm/dt) = (\lambda \times \dot{m})/A_D \quad [W/m^2] \quad (3.42)$$

Where $\dot{m} = dm/dt$ is the body mass change per unit of time and equals the mass flow with time (g/s), λ is the heat of vaporisation of water or enthalpy of evaporation (2430 J/g at 30°C), also called the latent heat of sweat evaporation and A_D is the body surface area (m²).

The evaporative cooling potential (E_{mass}) can be calculated as the latent heat content of the moisture that is evaporating from the overall human–clothing system as measured by the mass loss rate on the Sartorius scale in a steady-state condition (Havenith et al., 2008b, 2013):

$$E_{mass} = (\Delta Mass/dt) \times \lambda \quad [W/m^2] \quad (3.43)$$

There is an indication that the method does not necessarily represent the evaporative heat loss component from the skin, especially in non-isothermal conditions, where the heat losses from the skin are higher than those obtained by this method. The mass loss based vapour resistances and isothermal heat loss based vapour resistances can be miscalculated due to condensation occurring in low permeability clothing, which can lead to substantial errors in heat stress evaluation. The latent heat of evaporation is less than generally assumed, and the evaporative heat loss depends on processes on the body's surface as well as the processes in the clothing's microclimate (Havenith et al., 2008a).

The common assumption is that only moisture vapour actually leaving the clothing ensemble contributes to body cooling. Accordingly, the evaporative heat loss is calculated from the combined mass change of the body and the clothing rather than from the mass change of the body alone (Havenith et al., 2008a).

According to ISO 9886:2004, the gross body mass loss (Δm_g) of a person during a given time interval is the difference between the body masses measured at the beginning and at the end of this interval (ISO 9886:2004):

$$\Delta m_g = \Delta m_{sw} + \Delta m_{res} + \Delta m_o + \Delta m_{wat} + \Delta m_{sol} + \Delta m_{clo} \quad [g] \quad (3.44)$$

Where Δm_{sw} is mass loss due to sweat loss during the time interval (g), Δm_{res} is mass loss due to evaporation in the respiratory tract (g), Δm_o is mass loss due to the mass difference between carbon dioxide and oxygen (g), Δm_{wat} is mass variation of the body due to intake and excretion of water (g), Δm_{sol} is mass variation of the body due to intake and excretions of solids (g), and Δm_{clo} is mass variation due to variation of clothing or to sweat accumulation in the clothing (g).

The evaporative heat loss was traditionally determined only from all mass that was lost from the clothed person and corrected for metabolic and respiratory mass changes during the human wear test. This method was proven less precise because substantial errors may occur, which leads to overestimations of evaporative heat loss. Heat loss by evaporation cannot be estimated from the mass loss of the clothed person. During the low ambient temperatures, this is the after effect of the incomplete withdrawal of moisture from the skin, leading to the decrease in evaporative cooling efficiency (not all latent heat for the observed mass loss is taken from the skin), and an increase in the microclimate heat pipe effect without losing moisture. It is not due to the fabric conductivity in relation to the increase in the moisture content. In ambient temperatures below skin temperature, the microclimate heat pipe effect will transfer latent heat from the skin to the clothing without losing moisture, and there will be no obvious weight loss of the clothing (Havenith et al., 2008b). One can ascertain either the decrease in water vapour in the clothing by weighing the manikin-clothing assembly, the amount of water fed to the manikin, or the increase in heat loss of the wet manikin compared to dry in isothermal conditions to minimise the dry heat loss component and to determine the pure evaporative component of the heat balance (Havenith et al., 2008b). It can also be estimated using tables, where the data was obtained due to a sweating thermal manikin in a static condition and in a standing posture in low air movement with an air velocity below 0.2 m/s. There is a fourth method of estimation based on the relationship between the clothing water vapour resistance and the thermal insulation value of the clothing (dry heat resistance) (ISO 9920:2008; Havenith et al., 2008a).

The vaporised moisture leaving the clothing contributes to the body cooling and the evaporative heat loss. The wicking of sweat from the skin may have a positive effect when the skin is not fully wet by increasing the surface area of evaporation. However, the wicking of moisture into the clothing layers before it evaporates may reduce the body cooling effect and will lower the total available cooling power (Havenith et al., 2008b, 2013).

The evaporative cooling efficiency or the effective latent heat of evaporation (η_{app}) can be calculated (Havenith et al., 2013):

$$\eta_{app} = E_{app}/E_{mass} \quad [\varnothing] \tag{3.45}$$

The apparent evaporative cooling efficiency (η_{app}) is the apparent evaporative heat loss of the wet manikin (E_{app}) divided by the evaporative cooling potential (E_{mass}) under the same temperature condition (Havenith et al., 2008b). The real evaporative cooling efficiency of the body (η_{real}) is the real evaporative heat loss from the skin at 34°C (E_{real}) divided by the evaporative cooling potential (E_{mass}) in a given condition. If the apparent evaporative cooling efficiency (η_{app}) is higher than the real evaporative cooling efficiency of the body (η_{real}), the moisture evaporation is not the only mechanisms of the moisture removal. The apparent evaporative heat loss is lower

Modelling Heat Losses from the Human Body

due to the mass loss for the higher temperatures, especially for the impermeable fabrics. However, at lower temperatures, it becomes even higher, especially with the lower permeability of the clothing due to the evaporative cooling potential. When clothing gets wet and/or sweat migrates into the clothing, the evaporative cooling efficiency of body decreases (less cooling is provided to the body per gram of evaporated sweat/moisture) (Havenith et al., 2008b).

One can also calculate the effective latent heat of evaporation (λ_{eff}) in order to compare the latent heat of evaporation that benefits the body when clothing is worn. It is the amount of energy taken from the body for the evaporation of a given quantity of sweat (Havenith et al., 2013):

$$\lambda_{eff} = E_{app}/(\Delta \text{Mass}/dt) = \eta_{app} \times \lambda \quad [J/g] \quad (3.46)$$

The effective latent heat of evaporation (λ_{eff}) is dependent on the distance to the skin, the clothing permeability and on the ambient temperature. A substantial decrease in the value of the effective latent heat of evaporation is present when moisture is wicked away from the skin before it evaporates (Rossi et al., 2004; Havenith et al., 2008b, 2013).

The heat production forces the local temperature to increase, thereby decreasing the temperature gradient from the skin to the condensation spot and increasing the gradient between that spot and the environment (Lotens et al., 1990). The values of the effective latent heat of evaporation (λ_{eff}) and the effective latent heat of evaporation (η_{app}) depend on the ambient temperature, the permeability of clothing and the distance from the source of the evaporation. The values of the effective latent heat of evaporation (λ_{eff}) and the effective latent heat of evaporation (η_{app}) for permeable clothing will have their lowest value when the ambient temperature is equal or above the skin temperature, meaning no condensation is taking place and will increase with lowering temperature. As for the distance from the source of the evaporation, the effective latent heat of evaporation (λ_{eff}) decreases by 11% for evaporation from the skin when underwear and a permeable coverall is worn. The λ_{eff} for evaporation from the underwear decreases by 28% with the permeable coverall as the outer layer. With evaporation in the outermost layer, a decrease of 62% for no under clothing and an increase of about 80% with more layers between the skin and the wet outerwear was proven. In semi-and impermeable outerwear, the added effect of condensation in the clothing opposes this effect (Havenith et al., 2013).

The apparent evaporative heat loss (E_{app}) is the increase in heat loss compared to dry when evaporation is present; in other words, the manikin skin is wet. It includes the heat loss due to the evaporation of sweat at the skin surface, wet conduction (conduction of moisture/sweat) and evaporation–condensation (Havenith et al., 2013):

$$E_{app} = E - (C + R) \quad [W/m^2] \quad (3.47)$$

It can be calculated as the overall heat loss of a wet manikin subtracted from the heat loss of a dry manikin at the same temperature. Real evaporative heat loss (E_{real}) is the total wet heat loss of the manikin, measured at a skin temperature of 34°C. It thus accounts only for the heat loss via evaporation (Havenith et al., 2008b).

The effective cooling of the body is less than expected due to the latent heat of the evaporated sweat. The lower the permeability of the clothing will generate the lower values of the real evaporative efficiency and apparent latent heat of evaporation. If the permeability of the clothing is lower, the real evaporative cooling efficiency (η_{real}) will be lower and more adsorption–desorption cycles of moisture will occur through the clothing due to the higher microclimate vapour pressures (Havenith et al., 2008b).

In clothing aimed to protect against the cold, the vapour permeability of the fabrics and clothing is also an extremely important factor. The moisture accumulation affects the cold protection after the reduction in activity level since accumulated moisture evaporates and substantially cools the person, causing an after chill and increasing the risk of hypothermia. Other important factors to bear in mind when designing the cold-protective clothing, apart from the vapour resistance/permeability index and thermal insulation, include the water tightness, the air permeability (affecting heat resistance in the wind) and wicking. The heat and vapour resistance of the fabric may decrease due to the fabric's air permeability as the wind speed increases. Lower fabric air permeability usually provides better wind protection in cold-protective clothing. The moisture produced by the human body will partly condense within the layers of an ensemble during the cold exposures. The moisture should be reduced in the fabric layers to avoid wetting and a simultaneous thermal insulation reduction. The material's resistance against water penetration (waterproofness) should be high in cold-protective clothing. It is obtained due to special applications during fabric production such as coating and membranes to reduce the vapour permeability and to increase the waterproofness and the windproofness. The wicking ability of the fabrics is important for the underwear and next-to-skin materials since such fabrics remove the moisture away from the skin and thus reduce the discomfort (Rossi et al., 2004; Havenith, 2009).

The change in conductivity of a clothing ensemble including air layers showed that less than 3% can be attributed to wet conduction, while the majority is attributed to the microclimate heat pipe effect at lower temperatures (the evaporation of moisture at the skin that re-condensates in clothing and releases heat to the outer clothing layers) (Richards, 2004; Havenith et al., 2008b). For impermeable clothing combinations, wet conduction can only explain a small fraction of the additional heat loss. Condensation of sweat within clothing occurs at locations where the saturated vapour pressure is reached but this presents a big problem for clothing under cold condition condensation since the inner surface of an impermeable outer clothing layer heats up by several degrees. Findings by Richards et al. suggest that some of the outer layer condensation drips back down to the skin and underwear layers and evaporates again and is then re-condensed on the outer layer in a cyclic manner. This cyclic process is presumed to account for the large increases in the outer layer temperatures and increase in the temperature gradient between the outer layer and the environment, which enables heat to be lost to the environment more efficiently. Under steady-state conditions, the additional heat loss occurs due to a cycle of evaporation and condensation of moisture. Under transient conditions, the additional heat loss occurs due to the sorption and desorption of moisture. Thus, depending on the permeability of the clothing, the total heat loss under cold conditions can be much higher than expected from the sum of dry and evaporative heat loss. This fact needs to be considered in any future standard, which considers heat transfer through clothing (Richards et al., 2008).

In the case of the heat transfer between the skin and Burton's location of condensation and the skin and the spot of evaporation are dependent on the insulation of the clothing layers between the skin and these spots and on the total insulation of the clothing plus the surface air layer (Havenith et al., 2013).

The combination, the permeability and the hydrophilicity of the fabric layers influence the water vapour transfer and condensation effects in multilayer textile combinations, which is of high importance for the prevention of the condensation at lower ambient temperatures (Rossi et al., 2004).

The locus of evaporation, called the relative evaporative locus (REL), is defined as (Rossi et al., 2004; Havenith et al., 2013):

$$\text{REL} = \left(I_\text{T} - I_\text{skin to wet layer}\right)/I_\text{T} = I_\text{wet layer to environmet}/I_\text{T} \tag{3.48}$$

Where $I_\text{wet layer to environment}$ is the insulation between wet layer and environment, $I_\text{skin to wet layer}$ the insulation between the skin and the wet layer, and I_T is the total dry insulation of clothing ensemble. The humidity rather than the insulation value of the outer layer of clothing affects the evaporation process since the low humidity values increase the evaporation rate. The clothing should be able to transfer moisture from the skin surface, through the inner side. Surface contact angles below 90° will facilitate the moisture transfer process. Since the moisture is transferred through processes like diffusion, wicking, sorption and evaporation, the hydrophobic materials may require some force, pressure or an opening to allow efficient skin-clothing moisture transfer (Li, 2001; Kamalha et al., 2013).

A transient thermal model proposed by Wu and Fan integrated human thermoregulation and heat and moisture transfer through clothing, including ventilation induced by body motion, liquid sweat movement and the coupling effects of heat and moisture transfer (Wan and Fan, 2008). Another study was performed by Rugh et al. to predict thermal comfort in a transient, non-uniform thermal environment due to a physiological model of the human thermal regulatory system, a physical model (manikin) of the human body including heating and sweating and an empirical model to predict local and global thermal sensation and comfort (Rugh et al., 2004).

A study by Keiser et al. showed that the moisture content of a single layer is dependent on the material properties, but mainly on properties of the neighbouring layers or even of the whole combination. The overall moisture distribution in multilayer protective clothing can be influenced by using different combinations of hydrophilic and hydrophobic textile layers since direct contact between a hydrophilic station uniform layer absorbs moisture and leads it to the outer layers, while a hydrophobic station uniform layer only takes up little moisture (Keiser et al., 2008).

Wu and Fan proved that by placing the hygroscopic battings in the inner region and non-hygroscopic battings in the outer region of clothing, assembly reduces moisture accumulation (condensation) and associated dry heat loss within clothing (Wu and Fan, 2008). Rossi et al. proved that water vapour resistance and condensation increases with decreasing temperatures for clothing systems containing laminates (Rossi et al., 2004). With the temperature values decreasing to 0°C or below, condensation will get more pronounced, and the weight gain of the clothing may become a good indicator of the condensation effect (Havenith et al., 2008b).

Multilayer fabrics are usually used for cold weather and thermal protective clothing. The inner layer provides next-to-skin comfort by wicking the sweat at the skin's surface for better evaporative cooling and faster drying, the middle layer provides insulation, while the outer layer protects against environmental conditions and allows water vapour transfer to the environment (Huang, 2016). The amount and distribution of moisture in a fabric changes the heat transfer, thus affecting its thermal protective performance. However, there are studies showing completely opposite findings on moisture effects on multilayer fabric systems under heat flux flame exposures. Some researchers observe the decrease of thermal protection of multilayer fabric systems with water uptake (Rossi and Zimmerli, 1995; Mäkinen et al., 1988; Barker et al., 2006), while others found that thermal protection of fabric combination increased with the increasing moisture content under radiant heat exposure without air gap (Lawson et al., 2004; Song, 2011; Wang et al., 2012). Wang et al.'s study, however, indicated that when the air gap is small enough or absent between the outer shell fabric and moisture barrier, the moisture increases the thermal protection performance of multilayer fabrics system (Wang et al., 2012).

REFERENCES

Advanced Characterization and Testing of Textiles. 2018. ed. P. Dolez, O. Vermeersch and V. Izquierdo, Cambridge, MA: Elsevier Ltd., ISBN 978-0-08-100453-1.

Angelova, R. A. 2016. *Textiles and Human Thermophysiological Comfort in the Indoor Environment.* Boca Raton, FL: CRC Press, ISBN 978-1-4987-1539-3.

ANSI/ASHRAE Standard 55-2017 Thermal Environmental Conditions for Human Occupancy. 2017. Atlanta, GA: American Society of Heating, Refrigerating and Air Conditioning Engineers Inc. (ASHRAE), ISSN 1041-2336.

ASHRAE Handbook of Fundamentals (SI ed.). 2017. Atlanta, GA: American Society of Heating, Refrigerating, and Air Conditioning Engineers Inc., ISBN 1931862702.

ASHRAE Thermal Environmental Conditions for Human Occupancy. 1992. Atlanta, GA: American Society of Heating, Refrigerating, and Air Conditioning Engineers Inc. (ASHRAE).

Atkins K. and M. Thompson. 2000. A spreadsheet for partitional calorimetry. *Sportscience* 4(3). https://www.sportsci.org/jour/0003/ka.html (accessed on May 5, 2018).

Balmer, R. T. 2011. *Modern Engineering Thermodynamics.* Philadelphia, PA: Elsevier Inc., ISBN: 978-0-12-374996-3.

Bánhidi, L., L. Garbai and I. Bartal. 2008. Research on static and dynamic heat balance of the human body. *WSEAS Transactions on Heat and Mass Transfer* 3(3): 187–197.

Barker, R. L., C. Guerth-Schacher, R. V. Grimes and H. Hamouda. 2006. Effects of moisture on the thermal protective performance of firefighter protective clothing in low-level radiant heat exposures. *Textile Research Journal* 76(1): 27–31.

Bartels, V. T. 2006. Physiological comfort of biofunctional textiles. *Current Problems in Dermatology.* 33: 51–66.

Bedek, G., F. Salaün, Z. Martinkovska, E. Devaux and D. Dupont. 2011. Evaluation of thermal and moisture management properties on knitted fabrics and comparison with a physiological model in warm conditions. *Applied Ergonomics* 42: 792–800.

Behera, B. K., S. M. Ishtiaque and S. Chand. 1997. Comfort properties of fabrics woven from ring-, rotor-, and friction-spun yarns. *Journal of The Textile Institute* 88(3): 255–264.

Behery, H. M. 2005. *Effect of Mechanical and Physical Properties on Fabric Hand.* Cambridge, UK: Woodhead Publishing.

Bejan, A. 2004. Flows in porous media, Chapter 2. In: *Porous and Complex Flow Structures in Modern Technologies*, ed. A. Bejan, I. Dincer, S. Lorente, A. Miguel and H. Reis. New York, NY: Springer-Verlag, 31–66, ISBN: 978-0-387-20225-9.

Belding, H. S., H. D. Russell, R. C. Darling and G. E. Folk. 1947. Analysis of factors concerned in maintaining energy balance for dressed men in extreme cold; effects of activity on the protective value and comfort of an artic uniform. *American Journal of Physiology* 149: 223–239.

Berglund, L. G. 1998. Comfort and humidity. *ASHRAE Journal* 40(8): 35–41.

Bhatia, D. and U. Malhotra. 2016. Thermophysiological wear comfort of clothing: An overview. *Journal of Textile Science & Engineering* 6(2).

Bröde, P., K. Kuklane, G. Havenith. 2009. A thermal manikin study on the heat gain from infrared radiation with wet clothing. In: *Proceedings of the 13th International Conference on Environmental Ergonomics, Environmental Ergonomics XIII.* Wollongong, Australia: University of Wollongong, ISBN: 978-1-61782-976-5.

BS EN 14058:2017 Protective Clothing – Garment for Protection Against Cool Environments. 2017. BSI.

Burton, A. C. and O. G. Edholm. 1955. *Man in a Cold Environment.* London, UK: Edward Arnold Ltd.

Dincer I. and M. A. Rosen. 2013. *Exergy: Energy, Environment and Sustainable Development*, 2nd edition. Boca Raton, FL: Elsevier Ltd., ISBN 978-0080-4452-98.

DuBois, E. F. 1939. Heat loss from the human body. Harvey Lecture. *The Bulletin* 143–173.

Epstein Y. and D. S. Moran. 2006. Thermal comfort and the heat stress indices. *Industrial Health* 44: 388–398.

Fabbri K. A brief history of thermal comfort: From effective temperature to adaptive thermal comfort, Chapter 2. 2015. In: K. Fabbri, *Indoor Thermal Comfort Perception.* Switzerland: Springer International Publishing.

Fanger, P. O. 1970. *Thermal Comfort: Analysis and Applications in Environmental Engineering.* Copenhagen, Denmark: Danish Technical Press.

Fanger, P. O. 1973a. Assessment of man's thermal comfort in practice. *British Journal of Industrial Medicine* 30: 313–324

Fanger, P.O. 1982. *Thermal Comfort.* Malabar, FL: Robert E. Krieger Publishing Company.

Fanger, P. O., L. Bánhidi, B. W. Olesen and G. Langkilde. 1980. Comfort limits for heated ceilings. *ASHRAE Transactions* 86: 141–156.

Farrington, R. B., J. P. Rugh, D. Bharathan and R. Burke. 2004. Use of a thermal manikin to evaluate human thermoregulatory responses in transient, non-uniform, thermal environments. *2004 SAE International.* https://www.nrel.gov/transportation/assets/pdfs/20 04_01_2345.pdf (accessed on April 15, 2018).

Fourt, L. and N. R. S. Hollies. 1969. *The Comfort and Function of Clothing*, Technical Report 69-74-CE, Textile Series Report 162. Natick, MA: Harris Research Laboratories, Division of Gillette Research Institute Inc. Contract DAAG 17-67-C-0139 for US Army Natick Laboratories.

Fourt, L. and N. R. S. Hollies. 1970. *Clothing: Comfort and Functions.* New York, NY: Marcel Decker Inc.

Frackiewicz-Kaczmarek, J., A. Psikuta, M.-A. Bueno and R. M. Rossi. 2015. Air gap thickness and contact area in undershirts with various moisture contents: Influence of garment fit, fabric structure and fiber composition. *Textile Research Journal* 82(14): 1405–1413.

Gagge, A. P. 1937. A new physiological variable associated with sensible and insensible perspiration. *American Journal of Physiology* 20(2): 277–287.

Gagge, A. P. and R. R. Gonzalez. 1996. Mechanisms of heat exchange: Biophysics and physiology. In: *Handbook of Physiology Sect 4: Environmental Physiology*, ed. M. J. Fregly and C. M. Biatteis. Bethesda, MD: The American Physiological Society, 45–84.

Gagge, A. P. and Y. Nishi. 1976. Physical indices of the thermal environment. *ASHRAE Journal* 18(1): 47–51.

Gagge, A. P. and Y. Nishi. 1977. Heat exchange between human skin surface and thermal environment, Chapter 5. In: *Handbook of Physiology, Supplement 26: Reactions to Environmental Agents*, ed. D. H. K. Lee, H. L. Falk and S. D. Murphy, 69–92. Rockville, MD: American Physiological Society, ISBN: 9780470650714.

Gagge, A. P., A. C. Burton and H. C. Bazzet. 1941. A practical system of units for the description of the heat exchange of man with his environment. *Science* 94: 428–430.

Gagge, A. P., J. A. J. Stolwijk and Y. Nishi. 1971. An effective temperature scale based on a simple model of human physiological regulatory response. *ASHRAE Transactions* 77(1): 247–262.

Gibson, P., C. Kendrick, D. Rivin and L. Sicuranza. 1995. An automated water vapor diffusion test method for fabrics, laminates, and films. *Journal of Coated Fabrics* 24: 322–345.

Hardy, J. D. 1937. The physical laws of heat loss from the human body. *Proceedings of the National Academy of Sciences of the United States of America* 23(12): 631–637.

Hardy, J. D. and E. F. DuBois. 1938. The technic of measuring radiation and convection. *Journal of Nutrition* 15(5): 461–475.

Hardy, J. D. and P. Soderstrom. 1938. Heat loss from the nude body and peripheral blood flow at temperatures of 22°C to 35°C, *The Journal of Nutrition* 16(5): 493–510.

Hatch. K. L. 1993. *Textile Science*, 1st edition. St. Paul, MN: West Publishing Company, ISBN 978-0314-9047-13.

Havenith. G. 2002. The interaction of clothing and thermoregulation. *Exogenous Dermatology* 1(5): 221–230.

Havenith, G. 2005. Thermal conditions measurement, Chapter 60. In: *Handbook of Human Factors and Ergonomics Methods*, ed. N. Stanton, A. Hedge, K. Brookhuis, E. Salas and H. Hendrics. Boca Raton, FL: CRC Press, 552–573. ISBN 0-415-28700-6.

Havenith, G. 2009. Laboratory assessment of cold weather clothing. In: *Textiles for Cold Weather Apparel, Woodhead Publishing in Textiles, number 93*, ed. J. Williams. Cambridge, UK: Woodhead Publishing, 217–243, ISBN 978-1-84569-411-1.

Havenith, G., P. Bröde, E. den Hartog, K. Kuklane, I. Holmér, R. M. Rossi, M. Richards, B. Farnworth and X. Wang. 2013. Evaporative cooling: Effective latent heat of evaporation in relation to evaporation distance from the skin. *Journal of Applied Physiology* 114: 778–785.

Havenith, G., R. Heus and W. A. Lotens. 1990. Resultant clothing insulation: A function of body movement, posture, wind, clothing fit and ensemble thickness. *Ergonomics* 33(1): 67–84.

Havenith, G., I. Holmér and K. Parsons. 2002. Personal factors in thermal comfort assessment: Clothing properties and metabolic heat production. *Energy and Buildings* 34: 581–591.

Havenith, G., M. G. M. Richards, X. Wang, P. Bröde, V. Candas, E. den Hartog, I. Holmér, K. Kuklane, H. Meinander and W. Nocker. 2008a. Use of clothing vapour resistance values derived from manikin mass losses or isothermal heat losses may cause severe under and over estimation of heat stress. In: *Proceedings of the 7th International Thermal Manikin and Modelling Meeting*. Coimbra, Portugal: University of Coimbra. https://www.adai.pt/7i3m/Documentos_online/papers/5.havenith_UK_7i3m.pdf (accessed on May 15, 2017).

Havenith, G., M. G. M. Richards, X. Wang, P. Bröde, V. Candas, E. den Hartog, I. Holmér, K. Kuklane, H. Meinander and W. Nocker. 2008b. Apparent latent heat of evaporation from clothing: Attenuation and 'heat pipe' effects. *Journal of the Applied Physiology*, 104: 142–149.

Heat & Mass Transfer in Textiles, 2nd edition. 2011. ed. A. K. Haghi. Montreal, Canada: World Scientific and Engineering Academy and Society. WSEAS Press, ISBN: 978-1-61804-025-1.

Heat exchange between animal and environment, Chapter 2. 1975. In: *Man and Animals in Hot Environments. Topics in Environmental Physiology and Medicine*, ed. D. L. Ingram and L. E. Mount. New York, NY: Springer-Verlag New York Inc., ISBN-13: 978-1-4613-9370-2.

Hensen, J. L. M. 1991. *On the Thermal Interaction of Building Structure and Heating and Ventilating System*. PhD thesis, Technische Universiteit Eindhoven.

Holmér, I. 1989. Recent trends in clothing physiology. *Scandinavian Journal of Environ Health* 15(1): 58–65.

Holmér, I. 1995. Protective clothing and heat stress. *Ergonomics* 38:166–82.

Holmér, I. 2006. Protective clothing in hot environments. *Industrial Health*, 44(3): 404–413.

Holmér, I. and S. Elnäs. 1981. Physiological evaluation of the resistance to evaporative heat transfer by clothing. *Ergonomics* 24:63–74.

Holmér, I., H. Nilsson, G. Havenith and K. Parsons. 1999. Clothing convective heat exchange proposal for improved prediction in standards and models. *Annals of Occupational Hygiene* 43(5): 329–337.

Huang, J. 2016. Review of heat and water vapor transfer through multilayer fabrics. *Textile Research Journal* 86(3): 325–336.

Huang, J. 2006a. Thermal parameters for assessing thermal properties of clothing. *Journal of Thermal Biology* 31: 461–466.

Huang. J. 2006b. Sweating guarded hot plate test method. *Polymer Testing* 25: 709–716.

Ingram, D. L. and L. E. Mount. 1975 *Man and Animals in Hot Environments. Topics in Environmental Physiology and Medicine*. New York, NY: Springer-Verlag New York Inc., ISBN-13: 978-1-4613-9370-2.

ISO 11092:2014 Physiological Effects – Measurement of Thermal and Water–Vapour Resistance under Steady–State Conditions (Sweating Guarded-Hotplate Test). 2014. ISO-International Organization for Standardization.

ISO 15831:2004 Clothing – Physiological Effects – Measurement of Thermal Insulation by Means of a Thermal Manikin. 2004. ISO-International Organization for Standardization.

ISO 7730:2005 Ergonomics of the Thermal Environment-Analytical Determination and Interpretation of Thermal Comfort Using Calculation of the PMV and PPD Indices and Local Thermal Comfort Criteria. 2005. ISO-International Organization for Standardization.

ISO 9886:2004 Ergonomics—Evaluation of thermal strain by physiological measurements. 2004. ISO-International Organization for Standardization.

ISO 9920:2008 Ergonomics of the Thermal Environment-Estimation of Thermal Insulation and Water Vapour Resistance of a Clothing Ensemble. 2008. ISO-International Organization for Standardization.

Kamalha, E., Y. Zeng, J. I. Mwasiagi and S. Kyatuheire. 2013. The comfort dimension: A review of perception in clothing. *Journal of Sensory Studies* 28: 423–444.

Kaplan, S. and A. Okur. 2008. The meaning and importance of clothing comfort: A case study for Turkey. *Journal of Sensory Studies* 23: 688–706.

Keiser, C., C. Becker and R. M. Rossi. 2008. Moisture transport and absorption in multilayer protective clothing fabrics. *Textile Research Journal* 78(7): 604–613.

Kirk, W. and S. M. Ibrahim. 1996. Fundamental relationship of fabric extensibility to anthropometric requirements and garment performance. *Textile Research Journal* 36(1): 37–47.

Lawson, L. K., E. M. Crown, M. Y. Ackerman and J. D. Dale. 2004. Moisture effects in heat transfer through clothing systems for wildland firefighters. *International Journal of Occupational Safety and Ergonomics (JOSE)* 10: 227–238.

Lee, J.-Y. and J.-W. Choi. 2004. Influences of clothing types on metabolic, thermal and subjective responses in a cool environment. *Journal of Thermal Biology* 29: 221–229.

Lenzuni, P., D. Freda and M. Del Gaudio. 2009. Classification of thermal environments for comfort assessment. *Annals of Occupational Hygiene* 53(4): 325–332.

Li Y. 2001. The science of clothing comfort. *Textile Progress* 31(1–2): 1–135.

Li, Y., A. M. Plante and B. V. Holcombe. 1995. Fiber hygroscopicity and perceptions of dampness. Part 2. Physical-mechanisms. *Textile Research Journal* 65: 316–324.

Lotens, W. A. 1989. The actual insulation of multilayer clothing. *Scandinavian Journal of Work and Environmental Health* 15(1): 66–75.

Lotens, W. A. and G. Havenith. 1989. *Calculation of Clothing Insulation and Vapour Resistance*, TNO report. Netherlands: Netherlands organization for applied research.

Lotens, W. A. and G. Havenith. 1992. A comprehensive clothing ensemble heat model. In: *Proceedings of the 5th International Conference on Environmental Ergonomics*. Maastricht, Netherlands.

Lotens, W. A., F.J. G. van de Linde and G. Havenith. 1990. *The Effect of Condensation in Clothing on Heat Transfer*, TNO report, TNO Institute for Perception, Netherlands

Mahnic Naglic, M. and S. Petrak. 2017. A method for body posture classification of three-dimensional body models in the sagittal plane. *Textile Research Journal* 1–17.

Mäkinen, H., J. Smolander and H. Vuorinen. 1988. Simulation of the effect of moisture content in underwear and on the skin surface on steam burns of fire fighters. In: *ASTM STP 989: Performance of Protective Clothing: Issues and Priorities for the 21st Century, 2nd Symposium*, ed. S. Z. Mansdorf, R. Sager and A. P. Nielsen, Philadelphia, PA: American Society for Testing and Materials (ASTM), 415–421.

Matusiak M. and W. Sybilska. 2015. Thermal resistance of fabrics vs. thermal insulation of clothing made of the fabrics. *The Journal of The Textile Institute*. doi: 10.1080/00405000.2015.1061789.

McCullough, E. A. and B. W. Jones. 1984. *A Comprehensive Data Base for Estimating Clothing Insulation*. Report Institute Environment Research, IER Technical Report 8401. Manhattan, KS: Kansas State University.

McLean J. A. and G. Tobin. 1987. *Animal and Human Calorimetry*. Cambridge, UK: Cambridge University Press.

Mert, E., S. Böhnisch, A. Psikuta, M.-A. Bueno and R. M. Rossi. 2016. Contribution of garment fit and style to thermal comfort at the lower body. *International Journal of Biometeoroly* 60: 1995–2004.

Mert, E., A. Psikuta, M.-A. Bueno and R. M Rossi. 2017. The effect of body postures on the distribution of air gap thickness and contact area. *International Journal of Biometeorology* 61: 363–375.

Morgan, C. and R. de Dear. 2003. Weather, clothing and thermal adaptation to indoor climate. *Climate Research* 24: 267–284.

Nicholson, G. P., J. T. Scales, R. P. Clark and M. L. de Calcina-Goff. 1999. A method for determining the heat transfer and water vapour permeability of patient support systems. *Medical Engineering & Physics* 21: 701–712.

Nielsen, R., B. W. Olesen and P. O. Fanger. 1985. Effect of physical activity and air velocity on the thermal insulation of clothing. *Ergonomics* 28(12): 1617–1631.

Nishi, Y. and A. P. Gagge. 1971a. Moisture permeation of clothing: A factor governing thermal equilibrium and comfort. *ASHRAE Transactions* 76: 137–145.

Nishi, Y. and A. P. Gagge. 1971b. Humid operative temperature: A biophysical index of thermal sensation and discomfort. *Journal of Physiology (Paris)* 63 (3): 365–368.

Parsons, K. C. 2000. Environmental ergonomics: A review of principles, methods and models. *Applied Ergonomics* 31: 581–594.

Parsons, K. C. 2001. The estimation of metabolic heat for use in the assessment of thermal comfort. In: *Proceedings of a Conference on Moving Thermal Comfort Standards into the 21st Century*, ed. K. McCartney. Windsor, UK, 301–308

Parsons, K. C., G. Havenith, I. Holmér, H. Nilsson, J. Malchaire. 1999. The effects of wind and human movement on the heat and vapour transfer properties of clothing. *Annals of Occupational Hygiene* 43(5): 347–352.

Petrak, S., M. Mahnic and D. Rogale. 2015. Impact of male body posture and shape on design and garment fit. *Fibres & Textiles in Eastern Europe* 23(6–114): 150–158.

Plante, A. M., B. V. Holcombe and L. G. Stephens. 1995. Fiber hygroscopicity and perceptions of dampness. Part 1. Subjective trials. *Textile Research Journal*, 65: 293–298.

Pontrelli, G. J. 1977. Partial analysis of comfort's gestalt. In: *Clothing Comfort: Interaction of Thermal, Ventilation, Construction and Assessment Factors*, ed N. R. S. Hollies and R. F. Goldman. Ann Arbor, MI: Ann Arbor Science Publishers Inc., 71–80.

Prek. M. 2006. Thermodynamical analysis of human thermal comfort. *Energy* 31: 732–743.

Psikuta, A., J. Frackiewicz-Kaczmarek, I. Frydrych and R. M. Rossi. 2015. Quantitative evaluation of air gap thickness and contact area between body and garment. *Textile Research Journal* 85(20): 2196–2207.

Psikuta, A., A. Mark and R. M. Rossi. 2016. The impact of the outer layers in multi-layer clothing systems on the distribution of the air gap thickness and contact area. In: *Proceedings of the 7th International Conference on 3D Body Scanning Technologies*. Lugano, Switzerland.

Richards, M. G. M. 2004. *Report on Material Measurements Thermprotect*. EU Report Contract No. G6RD-CT-2002-00846. Loughborough University.

Richards, M. G. M., R. Rossi, H. Meinander, P. Broede, V. Candas, E. den Hartog, I. Holmér, W. Nocker andG. Havenith. 2008. Dry and wet heat transfer through clothing dependent on the clothing properties under cold conditions. *International Journal of Occupational Safety and Ergonomics (JOSE)* 14(1): 69–76.

Rohles, F. H., J. E. Woods and R. G. Nevins. 1973. The influence of clothing and temperature on sedentary comfort, *ASHRAE Transactions* 79: 71–80.

Rossi, R. M. and T. Zimmerli. 1995. Performance of protective clothing. In: *ASTM STP 1237*, 5th volume, ed. James S. Johnson and S. Z. Mansdorf. Philadelphia, PA: American Society for Testing and Materials (ASTM).

Rossi, R. M., R. Gross and M. May. 2004. Water vapour transfer and condensation effects in multilayer textile combinations. *Textile Research Journal* 74(1–6): 1–6.

Rugh, J. P., R. B. Farrington, D. Bharathan, A. Vlahinos, R. Burke, C. Huizenga and H. Zhang. 2004. Predicting human thermal comfort in a transient nonuniform thermal environment. *European Journal of Applied Physiology* 92(6):721–727.

Scott, R. A. 2000. Fibres, textiles and materials for future military protective clothing. In: *Ergonomics of Protective Clothing, Proceedings of 1st European Conference on Protective Clothing*, ed. K. Kuklane and I. Holmér. Stockholm, Sweden: National Institute for Working Life, 79–86, ISBN 91-7045-559-7.

Shephard, R. J. 1987. Metabolic consequence of exercise in the cold. In: *Adaptive Physiology to Stressful Environments*, ed. S. Samueloff and M. K. Yousef. Boca Raton, FL: CRC Press, 35–50, ISBN: 9780849364587.

Slater, K. 1977. Comfort properties of textiles, *Textile Progress* 9(4): 1–70.

Slater, K. 1985. *Human Comfort*, Springfield, IL: Charles C. Thomas Publisher, ISBN: 9780398051280.

Song G. (ed). *Improving Comfort in Clothing*. 2011. . Cambridge, UK: Woodhead Publishing Ltd, ISBN 978-1-84569-539-2.

Takada, S., S. Hokoi and K. Nakazawa. 2004. Measurement of moisture conductivity of clothing. *Journal of the Human-Environmental System* 7(1): 29–34.

Tracy, C. R. 1972. Newton's law: Its application for expressing heat losses from homeotherms. *Bioscience* 22(11): 655–659.

Umbach, K. H. 1988. Physiological tests and evaluation models for the optimisation of performance of protective clothing. In: *Environmental Ergonomics*, ed. I. B. Mekjavić, E. W. Banister and J. B. Morrison. London, UK: Taylor and Francis, 139–161.

van Hoof, J., M. Mazej and J. L. M. Hensen. 2010. Thermal comfort: Research and practice. *Frontiers in Bioscience* 15(2): 765–788.

Varshney, R. K., V. K. Kothari and S. Dhamija. 2010. A study on thermophysiological comfort properties of fabrics in relation to constituent fibre fineness and cross-sectional shapes. *The Journal of The Textile Institute* 101(6): 495–505.

Vogelaere, P., J. Deklunder, J. Lecroart, G. Savourey and J. Bittel. 1992. Factors enhancing cardiac output in resting subjects during cold exposure in air environment. *The Journal of Sports Medicine and Physical Fitness* 32(4): 378–386.

Wan, X. and J. Fan. 2008. A transient thermal model of the human body–clothing–environment system. *Journal of Thermal Biology* 33: 87–97.

Wang, Y., Y. Lu, J. Li and J. Pan. 2012. Effects of air gap entrapped in multilayer fabrics and moisture on thermal protective performance. *Fibers and Polymers* 13(5): 647–652.

Wang, Z., R. de Dear, M. Luo, B. Lin, Y. He, A. Ghahramani and Y. Zhu. 2018. Individual difference in thermal comfort: A literature review, *Building and Environment*. doi:10.1016/j.buildenv.2018.04.040.

WHO. http://www.euro.who.int (accessed on July 20, 2017).

Wilson, C. A., R. M. Ling and B. E. Niven. 2000. Multiple bedding materials and the effect of air spaces on 'wet' thermal resistance of dry materials. *Journal of Human-Environment Systems* 4(1): 23–32.

Winslow, C. E., L. P. Herrington and A. P. Gagge. 1936a. The determination of radiation and convection exchanges by partitional calorimetry. *American Journal of Physiology* 116(3): 669–684.

Winslow, C. E., L. P. Herrington and A. P. Gagge. 1936b. A new method of partitional calorimetry. *American Journal of Physiology* 116(3): 641–655.

Winslow, C. E., L. P. Herrington and A. P. Gagge. 1937. Physiological reactions of the human body to varying environmental temperatures. *American Journal of Physiology* 120(1): 1–21.

Winslow, C. E., L. P. Herrington and A. P. Gagge. 1939. Physiological reactions and sensations of pleasantness under varying atmospheric conditions. *Transaction of American Society of Heating and Ventilation Engineers (ASHVE)* 44: 179–196.

Wong A. S. W. and Y. Li. 2002. Clothing sensory comfort and brand preference. In: *Proceedings of the 4th IFFTI International Conference*. Hong Kong, China, 1131–1135.

Wray, W. O. 1980. A simple procedure for assessing thermal comfort in passive solar heated building. *Solar Energy* 25: 327–333.

Wu, H. and J. Fan. 2008. Study of heat and moisture transfer within multi-layer clothing assemblies consisting of different types of battings. *International Journal of Thermal Sciences* 47: 641–647.

Yoo, W. S. and E. Kim. 2012. Wear trial assessment of layer structure effects on vapor permeability and condensation in a cold weather clothing ensemble. *Textile Research Journal* 82(11): 1079–1091.

Zhang, P., Y. Watanabe, S. H. Kim, H. Tokura and R. H. Gong. 2001. Thermoregulatory responses to different moisture-transfer rates of clothing materials during exercise. *Journal of the Textile Institute* 92(1–4): 373–379.

Zhang, X. H. and Y. Y. Wang. 2012. Effects of clothing ventilation openings on thermoregulatory responses during exercise. *Indian Journal of Fibre & Textile Research* 37: 162–171.

4 The Importance of Globally Accepted Test Methods and Standards

Clothing protects the human body from external hazards or unpleasant environments and generates a convenient microenvironment within a microclimate so the individual can have a state of well-being while wearing clothing enabling them to sustain work and properly respond to situations in harsh environmental conditions (Mert et al., 2017).

The most important function of globally accepted test methods is in the assessment of product properties in order to predict the performance for specific applications. The importance of test methods and standards for assessing the thermophysiological properties of textile and non-textile materials for clothing production and products related to thermal protection gets even more relevant. Clothing, besides being an elementary human need for maintaining health and protecting against environmental conditions, can tremendously improve sport and work performance. The different specialised variations of protective clothing serve as a barrier against dangerous agents for people working in hazardous environments, protecting them against thermal, chemical, biological, nuclear, mechanical, radiative and physical hazards.

Protective clothing is categorised according to its application as chemical protective clothing (chemical splash and vapour protection, dry chemical handling), biological (microbial) protective clothing (clean-room apparel, hospital textiles), nuclear protective clothing, ballistic, puncture- or cut-resistant protective clothing and thermal protective clothing (fire fighting or cold-protective clothing) (Bajaj and Sengupta, 1992; Shishoo, 2002).

The thermal protective clothing brings protection against convective heat exposures such as flames; contact heat; radiant heat sources; fire sparks, drops from melted substances and explosions (glass, metals, steel, munition), hot gases, vapours or liquids and against severe cold and frost (Zimmerli, 2000; Bajaj and Sengupta, 1992). Firefighters' suits are made from fire-retardant materials and especially used to protect against high temperatures, hot liquids and flames, Figure 4.1.

The main material properties to be tested in heat and flame protective clothing are thermal conductivity, flame retardancy, heat resistance, high reflectiveness against heat radiation and high insulation that does not place the wearer under heat stress but serves to protect against high temperatures. The cold-protective clothing protects the human body from the effects of cold temperatures, wind, snow and rain. Cold-protective clothing should be designed in a manner which prevents condensation of water vapour after prolonged use (Bajaj and Sengupta, 1992), Figure 4.2.

FIGURE 4.1 The basic layers of a firefighters clothing ensemble.

Firefighter garment testing and standards concentrate mostly on material properties and requirements (material heat and vapour resistance, flammability, reflection properties). Additional developing standards assess the risk of burn injuries using an instrumented manikin. The firefighter garment testing often neglects the effect of stitching, seams, treatments, clothing design, sizing and fit, as well the interaction between the clothing and additional gear (Havenith and Heus, 2000). The standardised testing methods for defining the thermal properties of clothing usually do not take into account human behaviour and how people interact within work situations and environments, so the adaptive approach to risk assessment should be used (Parsons, 2000b).

4.1 TESTING THE THERMAL PROPERTIES OF TEXTILES AND CLOTHING

The human response to environments can be assessed by four principal methods – objective measurement methods, subjective methods, behavioural methods and modelling methods (Parson, 2000a). The objective measuring methods provide direct measures of human metabolism and responses to the affecting factors imposed by different climatic conditions. The disadvantage is the need for a representative sample of the user population exposed to the selected environment. More often, the danger lies in the measuring instruments interference or misinterpretation of the results

Value of Globally Accepted Test Methods and Standards 85

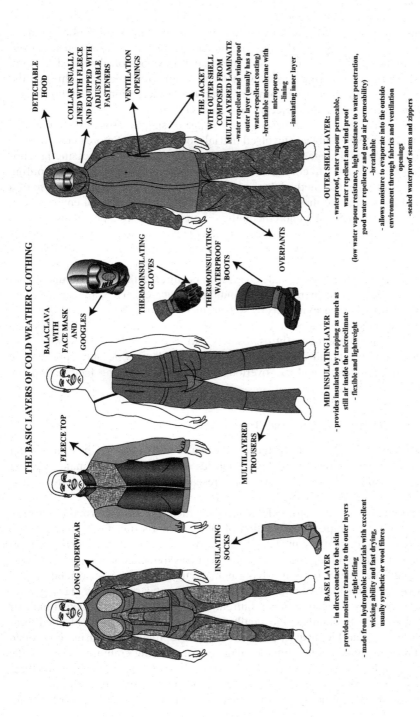

FIGURE 4.2 The basic layers of a cold-protective clothing ensemble.

and inability to predict subjective outcomes such as comfort (Parsons, 2000a). The clothing's moisture resistance and conductivity can be obtained using the empirical equations (Jürgens, 1924) found in the database developed by McCullough et al. (McCullough et al., 1989) by the wet cup method (ISO 12572:2016) and improved wet cup method (Takada et al., 2004). The insulation values of fibres and clothing can be determined in various ways by material weight and thickness, through the standard values and calculation formulas, and due to measurements by hot plates, thermal manikins or human subjects (Wang and Gao, 2014).

The subjective methods are based on the usage of simple rating scales to assess the thermal comfort of subjects involved in trials, or the usage of more detailed questionnaires. Their advantage lies in their relative applicability for assessing psychological responses related to comfort. However, they are difficult to design according to a number of potential methodological approaches and often require the use of a representative sample of the user population exposed to the selected environment. Another disadvantage is the unpredictability of answers due to a person not be able to detect the effects of the physiological strain and due to environmental stress interfering with a person's capacity to make a reliable subjective assessment (Parsons, 2000a).

The behavioural methods include changes in posture, clothing, environment conditions and adjustments, change in pace and the activity, etc. They are explained on the basis of the behavioural models by well-trained experts (Parsons, 2000a).

Using the subjective scales and environmental surveys allows for direct assessment of physical environments (https://oshwiki.eu). The modelling methods are easy to use for both design and evaluation, but those models provide only approximate responses by individuals and can overlook some of the affecting factors (Parson, 2000a).

4.1.1 Textile Thermal Comfort Testing

The dominating determinants of human thermal comfort are the heat and water vapour transfer properties of the textile fabrics (Huang, 2006). Climate chamber experiments and field studies are two major research methods for the study of thermal comfort (Wang et al., 2018).

The physical properties of textile materials, which affect the thermal comfort in humans involve a complex combination of heat and mass transfer, occurring both separately or simultaneously in steady-state or transient conditions. There are typically two types of clothing and fabric comfort measurements, laboratory and wear trial testing. The wearer trials involving human subjects can be performed in field tests or under controlled conditions. The environmental conditions show greater variability during field testing thus require more repetition and a sufficient number of human subjects to obtain a statistically significant result, so the controlled wear trials are more plausible, usually less variable and reproducible. During controlled wear trials performed inside climatic chambers, measuring equipment records the objective data for further analysis (Nielsen et al., 1985; Lotens and Havenith, 1989, 1991). However, wear trials are time-consuming and expensive. Laboratory test methods are most commonly used to determine different aspects of comfort with high reproducibility and less variability than those measured on humans (Nicholson et al., 1999). The advantage of laboratory test methods is seen through the cost reduction and smaller time consumption in comparison to classical wear trials. However, physical measurement due to laboratory testing does not directly measure comfort and has to be correlated with human perception in wearer trials (Dolez et al., 2018).

There are different test methods to explain the dynamics of water vapour transmission through fabrics due to the diffusion and permeation properties of laminates and textiles (Pye et al., 1976; O'Brien et al., 1986; ASTM F 372-73, 1976; Wehner et al., 1988; Gibson, 1994).

Quick commercially acceptable methods to affirm the thermal properties in clothing are Kawabata's KES-FB7 Thermolabo II system (Kawabata and Niwa, 1996) and the Alambeta apparatus (Hes and Dolezal, 1989).

Thermal absorptivity is an objective measurement of the warm-cool feeling (Bhatia and Malhotra, 2016). The Alambeta apparatus measures the warm-cool feeling, as one of the most important thermal handles of the textiles. This warm-cool feeling is non-dependent on the experimental conditions. It is only dependent on the fabric's thermal conductivity, the fabric density and the fabric's specific heat. The Alambeta apparatus measures the thickness of the sample, the thermal resistance and the thermal absorptivity (transient heat transfer), called the warm-cool feeling (Hes and Dolezal, 1989; Hes, 2004, 2008).

The thermal conductivity is measured by steady-state and transient (or non-steady-state) methods, each of them applicable only to a limited range of materials. The transient methods directly measure thermal diffusivity and thermal conductivity during the heating process, while the steady-state methods are used to measure the thermal conductivity when a thermal state reaches complete equilibrium for the selected sample (the constant temperature). Those methods apply Fourier's law of heat conduction to measure thermal conductivity. There are a few disadvantages with the steady-state methods such as a long time to reach the required equilibrium and the expensive apparatus. However, the steady-state thermal conductivity measurement technique is most commonly used due to its high accuracy. In contrast, the transient method for measuring thermal conductivity takes little time. The apparatus measuring using steady-state techniques are the guarded-hotplate, the hot cylinder, the heat flow metre (Yüksel, 2016). A few of them are constructed especially for fabric and textile material testing. The guarded-hotplate, also known as the Poensgen apparatus, is one of the most commonly used apparatus due to its steady-state measurements for textiles. It relies on a steady temperature difference over a known thickness of a specimen and its primary purpose is to control the heat flow through the material. The Gustafsson probe or the hot-disc method is designed to measure both thermal conductivity and thermal diffusivity under transient states and can be applied to fabrics with the sensor in the shape of a double spiral of nickel covered material (Yüksel, 2016).

Breathability can be defined in terms of the moisture vapour transmission rate (MVTR) and the resistance to evaporative heat flow. The breathability of the textile material is characterised by its ability to allow transpiration of the water vapour through the material. The higher the water vapour permeability of textile material, the more vaporised sweat can be transported away from the body, which in turn makes the clothing more comfortable for the wearer because less sweat accumulates on the body.

The main types of breathable textiles, which protect from outside moisture but allow water vapour to penetrate include (Kramar, 1989; Holmes, 2000; Sen, 2008; Chinta and Satish, 2014):

1. Densely woven fabrics with water-repellent treatment (100% cotton, micro-filaments)
2. Microporous film laminates and coatings (e.g. Gore-Tex®)
3. Non-poromeric hydrophilic film laminates and coatings
4. A combination of microporous coating and a hydrophilic top coat
5. Retroreflective microbeads used on fabrics
6. Smart breathable fabrics
7. Fabrics based on biomimetic

The waterproof-breathable fabrics transport water vapour effectively from the inside to the outside atmosphere. They simultaneously protect the human body from wind, rain and cold to maintain a constantly comfortable clothing microclimate under a range of environmental conditions (Ahn et al., 2010).

The basic requirements for waterproof-breathable fabrics are waterproofness and breathability, windproofness, abrasion and tear resistance, good adhesion of the membrane or film coating to the base textile, easy care, wash resistance and washability, lightness and packability, durability, flexibility and stretchability, and other specific functional properties such as flame retardancy, chemical protection, high visibility, stain resistance and oil resistance. The waterproof-breathable fabrics are usually produced by coating or laminating the thin film over the basic textile material. Laminated waterproof-breathable fabrics are produced faster than the coated waterproof-breathable fabrics; however, the coated technology is less expensive and has a better handle than (lightweight) laminated fabrics, providing high flexibility. In opposition, a wider range of textile substrates can be laminated, and better waterproofness can be achieved and maintained after several washing cycles in laminated waterproof-breathable fabrics (Kramar, 1989).

The water vapour permeability is measured by three methods, the permetest method, cup method and MVTR cell method. The different terms are used to express the water vapour permeability of material for each method, so the results obtained from the different methods are not always comparable due to the different testing conditions and the units (Bhatia and Malhotra, 2016).

The international standard describing the measurement of the water vapour permeability of textiles is ISO 15496 (ISO 15496:2018). The other national standards to describe the water vapour transmission rate (WTR) are BS 7209 (BS 7209:1990) and ASTM E 96 (ASTM E 96, 2016). The upright cup method is used for water permeable textiles, while the inverted cup method is used for water impermeable textiles. Another way of determining the water vapour permeability of textiles is the sweating guarded-hotplate method (ISO 11092:2014). The disadvantage to the cup method is the complexity of the testing procedure, higher test costs, longer testing times and the need for well-trained laboratory staff (Hohenstein, press release). The American standard to describe this method currently in usage is the ASTM F 1868-17, which superseded the old ASTM F 1868-02.

The basic terms to be considered are (ASTM E 96, 2016; ISO 15496:2018):

- The water vapour permeability (defined as the property of the material, WVP) is the time rate of water vapour transmission through unit area of flat material of unit thickness induced by unit vapour pressure difference between two specific surfaces, under specified temperature and humidity conditions (ASTM E96) or the amount of water vapour diffusing through the textile per square metre, per hour and per difference of water pressure across the textile (ISO 15496:2018).
- The water vapour permeance (performance property, W_d) is the time rate of water vapour transmission through unit area of flat material or construction induced by the unit vapour pressure difference between two specific surfaces, under specified temperature and humidity conditions.
- The WTR is the steady water vapour flow in unit time through unit area of a body, normal to specific parallel surfaces, under specified temperature and humidity conditions at each surface.

The permetest skin model can measure the thermal and water vapour resistance of the fabric samples according to ISO 9920 and BS 7209. The apparatus measuring head is covered with the semi-permeable foil, while the test sample is placed on the curved moistened porous surface and exposed to the parallel airflow of adjustable velocity. The amount of evaporated heat is measured, while the water vapour resistance and relative water vapour permeability can be calculated (Hes, 2008).

Takada et al. proposed the wet cup method for fabric thermal conductivity measurements where the fabric specimen is placed between two styrene boards and tightly fixed to a box filled with the water according to ASTM E 96. It measures the moisture vapour transmission rate by directly determining the weight loss, with evaporation time (24 hours) of water contained in a cup, the top of which is covered by the cover ring. Afterwards, the water vapour permeability index is calculated. However, the wet cup method is difficult to apply to fabric conductivity measurements because it involves determination of the surface moisture transfer resistance at both sides of the sample and instability of measuring protocol at high air humidity (Takada et al., 2004; Das and Kothari, 2012; Bhatia and Malhotra, 2016).

The MVTR cell is a moisture vapour transmission cell, which measures humidity generated under controlled conditions as a function of time, and the change in humidity at a time interval gives the moisture transmission rate of the fabric according to ASTM F 2298 and EN ISO 12572:2001 (Bhatia and Malhotra, 2016; Nazaré and Madrzykowski, 2015; ASTM F 2298-03(2009)e1, 2009).. This is a gravimetric method and can be divided between a dry cup method, where a sample is placed inside the cup with a regulated source of outside constant humidity, and wet cup method, where the sample is placed inside cup filled with the liquid (pure water or saturated salt solution) (Bhatia and Malhotra, 2016; Nazaré and Madrzykowski, 2015; ASTM F 2298-03(2009)e1, 2009; Das and Kothari, 2012; www.mercer-instruments.com/en/application/MVRT-WVRT-permaence-DVS.html).

Liquid-moisture transfer through clothing can measure either the wetting or the wicking. The wettability of textile materials is measured by tensiometry or goniometry. The tensiometry uses the processor tensiometer to measure the wettability by measuring the wetting force due to the Wilhelmy method. The goniometry measures the wettability by measuring the contact angle between the liquid and the fabric due to the contact angle tester or drop analyser (Bhatia and Malhotra, 2016). The wicking through a textile material is explained by the amount of water wicked or the surface-water transport rate. The wicking can be determined by British BS 3424-12 for coated fabrics with very slow wicking properties (BS 3424-12:1996) or German DIN 53924:1997-03 with shorter duration (DIN 53924:1997-03). The BS 3424-12:1996 has been replaced by BS EN 12280-1:1998 (BS EN 12280-1:1998) and BS EN 12280-3:2002 (BS EN 12280-3:2002). The revised corresponding international standard used is ISO 1419 (ISO, 1419:2018).

The Hot Disc Method uses a thin disc-shaped sensor (radius 6.403 mm) placed between the two sample fabrics to measure the thermal conductivity within 20s, with an output power of 40 mW. The samples should be cut to 100×100 mm and compressed at 2 kPa (Bedek et al., 2011).

The dry heat transmission through the textile materials in a steady-state can be measured by (Kothari and Bal, 2005; Bhatia and Malhotra, 2016):

1. The cooling method where the rate of the human body cooling is determined after placing the fabric around the hot body under air current. The cooling method uses a hot body, surrounded by fabric whose outer surface is exposed to air and the rate for the cooling of the body is determined.
2. The disc method where the rate of the dry heat flow, also known as thermal transmissivity under particular conditions, is measured by placing the fabric between two metal plates of different temperatures (a heat source and a heat sink).
3. The constant temperature method involves placing the fabric on top of the isothermal hot body to measure the heat required to maintain the body at the constant temperature and afterwards calculating the thermal insulation value of the textile material. An isothermal hot body is insulated on all sides and energy required to maintain the hot body at a constant temperature is measured.

There are three types of measuring apparatus working according to the constant temperature method (Kothari and Bal, 2005; Bhatia and Malhotra, 2016):

1. Hot cylinder method, where the textile materials are wrapped around a constant temperature cylinder, contained within another coaxial cylinder immersed in water.
2. Hot guarded semi-cylinder method, which is based on Newton's law of cooling and thus requires the fabric to be stitched by a seam as the previous method.
3. Guarded-hotplate method.

Resistance to dry heat transfer of the various textile materials via a guarded-hotplate can be obtained due to different standard testing methods such as ASTM D (1518) (ASTM D 1518-14, 2014), ASTM F (1868) (ASTM F 1868-17, 2017) and ISO 11092 (ISO 11092:2014). Resistance to evaporative heat transfer can also be measured via hot plate apparatus adjusted to simulate sweating. It can be obtained using the above-mentioned standard testing methods, ASTM F (1868) (ASTM F 1868-17, 2017) and ISO 11092 (ISO 11092:2014).

A guarded-hotplate placed in the appropriate climatic chamber has different custom hoods to provide either still air conditions, horizontal air flow or vertical air flow at different levels for fabric thermal resistance measurements. Sweating guarded hot plates also come with different hood additions to provide either horizontal or vertical airflow at different levels (McCullough, 2012).

The most common laboratory test methods for acquiring thermal and water vapour resistance are performed using the sweating guarded-hotplate. The sweating guarded-hotplate is able to simulate both heat and moisture transfer through the clothing layers under steady-state conditions (Nicholson et al., 1999; Huang, 2006).

The sample is placed in the centre of the plate to transfer the heat or moisture upward along the specimen's thickness direction. Bottom and lateral guard heaters prevent heat loss from the specimen. The lateral guard, surrounding the hotplate, prevents lateral heat leakage from the edges of the specimen. The bottom heater prevents downward heat loss from the test section and the guard heater section (Fan et al., 2003; Huang, 2006).

The textile materials in direct contact with the skin can absorb the sweat from the skin and spread it to a larger area on the fabric to increase the heat loss due to evaporation (Takada et al., 1999). It is extremely important to know the water vapour and wicking properties of textile materials in order to make the appropriate selection while designing.

The water vapour permeability of polymer membranes laminated to textiles, which serve as a liquid barrier but is vapour permeable, are measured according to ISO 2528. Huang presented a new measuring apparatus to determine the vapour permeability of polymer membranes called the moisture permeation cylinder apparatus. This method is less time-consuming and cheaper compared to other standardised apparatus and allows smaller sample sizes (ISO 2528:2017; Huang, 2007). Since the during traditional sweating guarded-hotplate measurements, the test sample is directly in contact with the artificial skin, and the newly proposed method is designed to create a microclimate between the sample and the artificial skin and the test condition is closer to reality. This method involves placing a piece of waterproof and vapour permeable PTFE membrane laminated to a nylon tricot just above the water, simulating the body's skin. The sample is positioned 60 mm above the membrane in order to create the microclimate between the artificial skin and the sample (Zhang et al., 2010).

4.1.2 Clothing Thermal Comfort Testing

Never have studies employed a combination of thermal sensors directly measuring the total and radiative heat flux together, with the thermal manikin in different postures to determine the radiative and the convective heat transfer coefficients of the entire human body under natural convection, driven by the difference between the air temperature and mean skin temperature corrected using the convective heat transfer area. The convective heat transfer coefficient increases, while the radiative heat transfer coefficient decreases, in accordance with the temperature difference between the air and skin, corrected for the convective heat transfer area. The radiative and convective heat transfer coefficient depends on body postures having the greatest values while standing and the lowest values for the cross-legged sitting postures (Kurazumia et al., 2008).

Generally, radiative and convective heat flux released from the human body to the environment represents the majority of released heat. Around 40% of heat is released by convection and around 60% by radiation, if one does not take into account the evaporation and respiration (Murakami et al., 1999).

The factor of thermal conduction was almost entirely eliminated from the studies on human subjects in order to deduce convective and radiative heat transfer via convection and conduction applying Newton's laws on heat exchange (Hardy, 1937).

The thermal insulation of clothing, which corresponds to the thermal resistance of the textile materials and the clothing water vapour resistance, can be measured by different methods (Lotens and Havenith, 1989, 1991):

- Human wear trials in real or simulated environments (in field tests or under controlled conditions).
- Thermal manikin measurements.
- Regression due to the previously reported insulation values written in the tables, which have been obtained by static manikin under no air motion. The water vapour resistance tables are rarely available.
- The thermal insulation obtained by the regression due to the skin area covered with clothing, the thickness and the clothing weight. The water vapour resistance tables are rarely available.
- The calculation of heat and mass transfer when the geometry of the clothing and the body is known. The human body parts tend to be presented as cylinders.

Thermal manikin measurements are costly, time-consuming and only a limited number of sample values are available (Havenith et al., 2002), but they do provide a good basis for measuring clothing's thermal insulation. Since the thermal manikin measures heat loss in a relevant, reliable and accurate way, the method is quick, easily standardised and repeatable (Nilsson, 2003, 2004). Thermal manikins provide a rapid, accurate and reproducible simulation of the physical processes of heat loss to the environment (Wyon, 1989).

Thermal manikins were created with the intention of designing and modelling protective clothing insulation for specific conditions, e.g. military clothing and clothing for divers. The first manikin was a one segment copper manikin made for the U.S. Army in the early 1940s. The U.S. military set the trend of using the thermal manikins in clothing research by measuring the insulation of combat clothing ensembles during simulated adverse environmental conditions. At present, thermal manikins have a broader use, i.e. to assess the effect of clothing on a human body, to assess its influence on thermal comfort during work in a given clothing ensemble and to test innovative solutions bringing about a reduction in the thermal heat load (Endrusick et al., 2001; Bogdan and Zwolińska, 2012). But there is also another reason to use manikins when estimating thermal comfort. By definition from both ASHRAE 55 and ISO 7730, thermal comfort is *'that condition of mind that expresses satisfaction with the thermal environment'* (ANSI/ASHRAE Standard 55-2017; ISO 7730:2005).

In the beginning, thermal manikins were developed as a result of the global conflict in climatic extremes. Belding at the Harvard Fatigue Laboratory, Cambridge, used the first heated manikin to determine the comfort range of sleeping bags and Arctic uniforms in 1942. In 1960, research began to focus on the sweating thermal manikin, while in the 1970s thermal manikin studies started combining with human wear trials (Endrusick et al., 2001).

The performance features and application of thermal manikin measurement is seen through (Holmér, 1999):

1. Relevant simulation of human body heat exchange
2. Whole body and local heat exchange
3. Measurement of 3D heat exchange
4. Integration of dry heat losses in a realistic manner
5. Objective method for measurement of clothing thermal insulation
6. Quick, accurate and repeatable measurements
7. Cost-effective instrument for comparative measurements and product development
8. Provide values for prediction models
9. Measurements of the clothing insulation and evaporative resistance
10. Measurements of the heat losses

In the future, a significant increase in the implementation and in the acceptance of thermal manikins in the textile and apparel industry is expected (Pamuk, 2008).

For thermal insulation calculation methods, the serial, the global and the parallel methods can be used (Oliveira et al., 2008).

Thermal manikin measurements of clothing insulation and water vapour resistance show better reproducibility when compared to the human wear trials. However, the apparatus is costly, and the activities are restricted according to the performance capability provided by the apparatus, especially due to the mobility of the manikin (Lotens and Havenith, 1991).

There are different variations of the thermal manikin apparatus according to their usage. Resistance to dry heat transfer, known as the insulation value provided by clothing systems, is tested according to various methods, and some of them are ASTM F (1291) (ASTM F 1291-16, 2016), ASTM F 2732 (ASTM F 2370-16, 2016) and ISO 15831 (ISO 15831:2004). The sweating thermal manikin in an environmental chamber is used to measure the resistance to evaporative heat transfer provided by clothing systems according to ASTM F 2370 (ASTM F 2370-16, 2016) and the average cooling rate and duration of cooling provided by personal cooling systems worn with protective clothing according the ASTM F 2371 (ASTM F 2371-16, 2016).

Current thermal manikin testing can be done with the manikin moving, standing or sitting (McCullough, 2012).

Physiological wear comfort is usually obtained in three steps, first being the laboratory measurement, second being the predictive model creation and third being the wear comfort testing in practice. In assessing the thermophysiological impact and wear comfort of garments, it is insufficient to just perform the laboratory measurement with the thermal manikins. The necessity is imposed to create a predictive model, which translates the laboratory measurement data into wear comfort validated in practice by wear comfort testing (Bartels and Umbach, 1999).

The studies on convective heat exchange from the human body have focused on the convective heat transfer coefficient, which can be calculated in different ways. One method is to calculate the convective heat transfer coefficient on the basis of measured local coefficients and then to calculate the ratio of that area to the entire body surface area. Such calculations include non-convective heat transfer areas adding to less accuracy (Mochida, 1977; de Dear et al., 1997; Oguro et al., 2002).

In an experiment by de Dear et al., a thermal manikin with controlled skin temperature was used to deduce sensible heat transfer by convection and radiation from the human body under different indoor and outdoor climatic conditions. The thermal manikin was covered in a low-emissivity and highly reflective film, which enabled differentiating between the radiative and convective component of the dry heat flux. The limbs (hands, feet and arms) had higher convective heat transfer coefficients than the trunk, while the head and neck showed the smallest radiative and convective heat transfer coefficient. The lower parts of the body (lower legs and feet) had higher convective heat losses due to the wind influence (de Dear et al., 1997).

The regression models for the radiative and convective heat transfer coefficient are nowadays corrected for airspeed and physical activity (Havenith et al., 1990).

Other calculations estimated the total heat transfer from the body using physiological measurements and heat flow sensors, or through the standard formulas combined with the radiated heat to calculate the convective heat transfer coefficient of the body. All of the above-mentioned methods based their assumption on that convective heat exchange occurs uniformly across the whole-body surface whether or not the airflow is uniform or mixed and whether there is the natural convection or the forced convection. Since heat transfer through clothing is affected by body and air movements, the intrinsic clothing insulation (I_{cl}), the thermal insulation of surface air layer (I_a) and the total thermal insulation of clothing with a surface air layer (I_T) are defined for standard conditions, but are likely to change with air movement and body motion due to the increased convection in and on the surface of clothing layers. The dynamic heat transfer through clothing is reduced in comparison to the static heat transfer (Havenith et al., 2002).

Also, as Lotens and Haveith proved, there is a relationship between the intrinsic insulation of clothing (I_{cl}), the tightness of fit and the surface area factor (f_{cl}), since the loose-fitting garments must have a larger surface area factor than tight-fitting clothing of the same intrinsic insulation value due to the conductivity of all layers together (Lotens and Havenith, 1989).

The dry thermal insulation of the clothing is usually determined by dry thermal manikins and to a lesser extent by a sweating manikin. There are relatively few sweating manikins with the ability to measure the evaporative resistance of clothing, and the test procedures have not been standardised. Sweating manikins for obtaining wet heat transfer are categorised into three types, pre-wetted skin on dry manikins, manikins with sweat glands with a regulated constant water supply to the skin surface and the sweating fabric manikins (Gao et al., 2006). Most studies presented the clothing total insulation value, as dry heat transfer affects the factors buffering the heat transfer between the body and the environment, through modelling the combination of convection, radiation and conduction (Burton and Edholm, 1955; Holmér et al., 1999).

These sweating thermal manikins and numerical models of the human thermoregulatory system are used to evaluate the thermal response of an individual to transient, non-uniform thermal environments (Farrington et al., 2004). The method uses the evaporative mass loss to determine if the pure evaporative component of the heat balance is inadequate in all of the environmental conditions. The evaporation mass loss rate of moisture leaving the manikin's sweating surface shouldn't include any sweat dripping from the surface, nor any sweat absorbed by the clothing and is multiplied by the heat of vaporisation of the sweat at the measured surface temperature (ISO 9920:2007). There is a miscalculation in obtaining the values of the water vapour resistance measured by the sweating thermal manikin because the calculations involve mass loss. This method can be optimal for steady-state and clothing with normal permeability, but problems arise with reaching the stable measurements for less permeable clothing. The steady-state mass loss can take a long time to reach for less permeable clothing. There is also an apparatus limit, which should be taken into account. One is that the sweating manikin should be warmed before wetting but should also have sufficient power to quickly reach the steady-state after being wetted. The other is the regulation of the sweating and the problem with sweating dripping, which should be avoided. The more up-to-date and complex types of the sweating manikins are equipped with contemporary water supply systems, wicking skin and sweat glands to allow different sweat rates for different body parts (ISO 9920:2007).

The studies on convective heat transfer mostly involved testing on nude subjects because of the difficult separation of the convection and radiation components of heat transferred through clothing (Holmér et al., 1999).

The subject's movements and the wind force the saturated warm air through vents and the clothing layers. The heat transfer between the body and the environment depends on the relative or net air velocity, not the absolute airspeed. The corrections for the static insulation and the convective heat exchange values should be made for wind and walking (Havenith et al., 1990). However, correction due to walking takes into account only the effects of the movement and not those of the relative air velocity. The walking and the wind will decrease the clothing insulation values and increase the vapour permeability, thus allowing for greater heat transfer by convection and evaporation. The walking speed of pumping is calculated as the relative air velocity due to movement based upon metabolic rate and corrections have been made for insulation values presenting the static and dynamic conditions (McCullough and Kim, 1996; Holmér et al., 1999; Parsons et al., 1999).

The body motion effects the enclosed and surrounding air layers, while the wind mainly affects the surrounding air layer and the layer under the outer garment.

The combination of body movements, such as walking, and air velocity reduces the intrinsic clothing insulation (I_{cl}) and the surface air (I_a) thermal insulation when compared to static conditions (Gagge et al., 1941; Vogt et al., 1983; Nielsen et al., 1985; Havenith et al., 1990). Body movements and wind can compress the garments producing a decrease in air layer thickness. Body movements play a major role in intrinsic clothing thermal insulation reduction. The body movements force the air bellows effect, producing the circulation of air and pumping air inside the clothing or forcing exchange with the environment. The body movement also affects the outer surface air layer insulation (I_a), although the wind plays a major role in the reduction of the outer surface air layer insulation. Wind velocity up to 4 m/s does little to disturb the intrinsic clothing insulation (I_{cl}), due to the inability to penetrate through the materials and to disturb the inner air layers inside the clothing (Havenith et al., 1990). Nielsen et al. found a 30 to 36% reduction in the intrinsic thermal insulation of the clothing (I_{cl}) and a 20% reduction in the thermal insulation of the surface air (I_a) during cycling. The combined effect of wind and movement affect the clothing insulation and vapour permeability to a larger extent, reducing the clothing insulation up to 60% and more profoundly the water vapour resistance (Nielsen et al., 1985; Havenith, 2002). Nielsen et al. observed a reduction in the intrinsic thermal insulation value of clothing (I_{cl}) by 6 to 18% for a seated position compared to the standing position due to the air compression caused by the body weight and clothing deformation around joints (Nielsen et al., 1985).

The local convective heat transfer rates between the clothing layer and the external air layer are strongly dependent on the transport phenomena in the microclimate, especially with natural convection being the prevailing transport mechanism in the microclimate, thus pointing to importance of the natural convection in the microclimate and simultaneous decrease in emissivity of the fabric inner surface (Mayor et al., 2015).

The values obtained due to thermal manikin measurements serve as direct input for mathematical models for the prediction of thermal responses. Mathematical simulation modelling of the human body falls in the scope of human thermophysiological modelling. In order to describe the human thermal response to various climatic conditions, a number of models have been developed during the past years (Nilsson, 2004).

The variables in the mathematical simulation modelling include (Nilsson, 2004):

1. The number of details provided about the temperature field within the body
2. The way thermoregulatory responses are handled
3. The treatment of garments and boundary conditions
4. The possibilities to correlate with relevant measured data
5. The amount of computational effort and usability

The factors limiting the validity of mathematical models include (Nilsson, 2004):

1. Low resolution for describing the body temperature field
2. Inadequate description of one or more aspects of thermoregulatory responses

3. Insufficient information about the clothing
4. Lack of agreement between computed and measured results
5. The use of specialised computers and long calculation times

Thermoregulatory mathematical models are however versatile tools for determining heat distribution and risk for an individual under a variety of conditions (Pisacane et al., 2007). Some of the known models are the Fiala model, which addresses transient conditions, and the UC Berkeley thermal comfort model, which focuses on cooling effects in warm environments. A number of other thermal comfort models have also been proposed in the past 20 years focusing on inhomogeneous environments (Cheng et al., 2012). However, the predictive models for long-term exposure to cold are quite limited and lack the information on human responses for extended periods of time. The thermal comfort models need further development (Xua et al., 2005).

4.1.3 Subjective Judgements and Wear Trials: What Do We Have to Say?

Clothing comfort assessment by human perception analysis provides a sound basis for the comparison of apparel garments under combined microclimates and human-activity levels. The comfort studies seek the wearing and wearer conditions that influence the difference in human sensations and the descriptors used by individuals to describe these sensations. The classic physiological approach proposed by the ASHRAE has been extended in the past to include perceptions of thermal acceptability under different conditions such as work stress and the reduced air conditioning in a work environment. Application of perception analysis techniques to a wide variety of human–clothing situations and over the wide range of indoor and outdoor climatic conditions has shown that perception analysis is a powerful tool for delineating how the different features of apparel or fabric construction affect the wearers. In order to develop the descriptors for various sensations and the intensity rating of any sensation experienced using the scale, repeated experiments were performed in which the participants were asked to describe the sensations they experienced. The protocol test sequence is also developed as the standard during laboratory or outdoor wearing studies. The comfort sensation under specific conditions should be included in a total comfort assessment. Replicability during human comfort testing is obtained due to the controlled environmental and metabolic-activity conditions during the protocol test sequence. The results can be used to judge the garment and fabric characteristics and differences. The differences in garment ratings for individual wearers provide a sound basis for minimising effects due to the differences in wearers and maximising major differences in garments (Hollies et al., 1979).

The cyclic changes in the intensity of metabolism are reflected through the human subjective variables such as rectal and skin temperature, the environmental variables such as the temperature and humidity in the spaces between the clothing and the subjective thermal and humidity sensations. The interaction of the whole clothing system with the body and the resulting effects on the subjective sensations was seemingly neglected in the past (Vokac et al., 1972).

Various attitude scaling techniques have been applied to measure the perception of clothing comfort. Gagge et al. employed the 4-point category scale with comfort levels to quantify comfort and obtain the ratings of subjective perceptions (Gagge et al., 1967).

The scales, in the form of questionnaires, were developed for thermal subjective assessment methods. These are perceptual (scales of perceptual judgements on personal thermal state), evaluation (scales of evaluative judgements on personal thermal state), preference (thermal preference scales), acceptability (personal acceptability statement forms) and tolerance scales (Havenith, 2005a).

Humans form their subjective attitudes according to a cognitive component, an affective component and a behavioural component. Those components of the subjective attitude are the basis behind psychological scaling involved in clothing comfort research. The cognitive component forms the opinion based on a person's beliefs or information about the object. The affective component consists of the person's feelings of like or dislike concerning the object, while the behavioural component describes the person's action tendencies or predisposition towards the object (Li, 2001).

Subjective judgement scales are used in clothing comfort research. There are four types of psychological scales. The nominal scales are used to differentiate the general subject's data such as gender, age and place of living. The ordinal scales are used to rank the various properties of the fabrics or garments. The interval scales are used to obtain perceptions of various attributes of clothing. They are the most used scales. There is also a fourth type of scale called the ratio scales, which are mainly used to process the data generated from physical instruments (Li, 2001).

The advantage of using subjective rating scales in thermal comfort testing is that they are relatively easy to use and can be applied when the contributing factors to responses are not fully known. The disadvantage is shown through choosing the precise wording at different scale levels due to a number of methodological problems since thermal comfort is a subjective quantity. The use of the subjective rating scales requires a representative sample of the user population, which is quite costly and demands much effort in order to get repeatable results (Nilsson, 2004).

Each of the scales has its own set of rules, which become more restrictive moving from nominal to ratio scales. The numbers in nominal scales are used to categorise objects as a label for a class category. All members of a class have the same number, and no two classes have the same number. Nominal numbers cannot be added, subtracted, multiplied or divided. The numbers or symbols in ordinal scales are used to rank objects according to their characteristics (using the mode, median, quartile and percentile) and their relative position in the characteristics due to non-parametric statistics. Those categories do not show the magnitude of the differences between the objects. The ordinal scales involve the ranking along the continuum of the characteristic being scaled and give the order of preference. They do not show how one rating is preferred to another, since there is no information about the interval between any two ratings (Li, 2001; Nilsson, 2004).

The interval or cardinal scale has equal units of measurement. They can be used to interpret the order of scale ratings but also the interval between any two ratings. The condition for conversion between ordinal and cardinal scales is that the data forms an approximately linear relationship between mean thermal vote (MTV) and equivalent temperature at the individual level. The number appointed to the interval scales rank objects as the numerical presentation of the distances on the scale obtained through all statistical methods. The units of measurement are not fixed and are arbitrary, which means the interval data shows both the relative position of objects and the magnitudes of differences between the objects on the measured characteristics (Li, 2001; Nilsson, 2004).

The numbers in the ratio scales are used to rank objects such that numerically equal distances on the scale represent equal distances of the characteristics measured and have a meaningful zero through any possible statistical method (Li, 2001).

The attitude scales, in the form of the rating scale, or a group of rating scales that measure single dimensions of attitude components, are used to validate the subject's attitude on some matter. The subject places an attribute of the rated object in one or several numerically ordered categories. There are two types of rating scales, non-comparative (no rating standard through the graphic rating or itemised non-comparative rating) and comparative (standardised by selecting the numbers with a fixed position on the scale) (Li, 2001).

There are several characteristics of the itemised rating scales to bear in mind when designing the appropriate scale to place an attribute on the rated object. First of all, the response is affected by the nature and degree of verbal category descriptions. Five to nine categories can be used to combine several scales for one score depending on the purpose of the scale and the knowledge and interest of the subject comparing attributes across objects. The usage of a balanced or an unbalanced set of categories depends on the type of information required, and the assumed distribution of attitudes since a balanced scale provides an equal number of favourable and unfavourable categories. The balance scales are dependent on the type of the number corresponding to the specific category. The balance scale composed of the odd numbers usually has a neutral point incorporated in the middle of the scale. Using the forced to indicate an attitude showed that the subject will usually mark the midpoint of a scale if unsure on the subject, which will distort measures of central tendency and variance, so it is recommended to use unforced scales (Li, 2001).

When studying comfort attributes of textile products, paired comparisons and rank order ratings mostly use simple comparative methods. During the paired comparison, the subjects are presented with two objects to make a selection according to some specific criterion. The paired comparisons are generally limited to one attribute or to a couple of products on multiple attributes. The ranking scores obtained through paired comparisons is essentially ordinal data, but they cannot provide the magnitude of the perceived differences between samples nor their relative positions in a context of all possible relevant samples beyond a particular experiment. During the rank order rating, the subjects discriminate between objects in a manner close to the actual shopping environment according to some specific criterion. The number of statistical analyses is limited because only ordinal data is the possible output with rank order rating (Li, 2001).

Many clothing comfort studies use semantic differential rating scales. They have a series of bipolar rating scales made of word pairs that may be opposites or with two poles, one extreme and one neutral. Each bipolar word corresponds to a certain number. The central value usually represents the neutral state between the two extremes. The data obtained by these scales is analysed either by aggregate analysis or profile analysis, in order to compare data for the same group of individuals (Li, 2001).

There are basically no tools for measuring a person's opinion on comfort status, and thus subjective wear trials are used to evaluate subjective opinions and real performances of clothing systems. The subjective coolness, dampness and comfort evaluations, physiological measurements such as skin temperature and sweat rate, microclimate temperature and relative humidity are measured during wear trials and correlated into thermal insulation values (Kaplan and Aslan, 2016).

Perception of the comfort performance of clothing is best studied in wear situations, and this is an important technique for clothing comfort research. Since perceived comfort is experienced by a wearer under given environmental conditions, depends mostly on the tactile and thermal-moisture properties of the fabrics, physiological validation via human wear trials always provides useful information for functional clothing designers and customers (Wang et al., 2003; Umbach, 1988). Wear trials in a climatic chamber are performed to obtain information on the in-use properties and function of a system of clothing on a large scale, in different activities and climatic conditions (Vokac et al., 1972).

There are many studies involving subjective evaluation techniques such as subjective wear trials and consumer research techniques to understand consumers' comfort perception. The sensory descriptors related to thermal comfort are used for wear trials of thermal comfort evaluation methods (Kaplan and Okur, 2008).

The best practice in clothing comfort research on the basis of the wear trial experimental technique can be adopted due to Hollies et al. study (Hollies, 1977).

The components of a wear trial experimental technique on the basis of Hollies et al. Study include (Hollies, 1977):

1. Generating descriptors with respondents
2. Selecting testing conditions to maximise the opportunities for the perception of various sensations
3. Designing attitude scales in the way of rating sheets to obtain various responses to particular garments
4. Conducting wear trials in controlled environmental chambers according to the predetermined protocol
5. Collecting data and analysing and interpreting the results

Li et al. used wear trials to deduce the relative importance of different components of comfort for fabrics used for different end uses and to calculate the total wear comfort index. This index combines different parameters responsible for thermophysiological and tactile comfort by weighting factors. The data from surveys and wear trials showed that the perceived comfort was different under various combinations of physical activities and environmental conditions (Li et al., 1991).

Wear trials have been extensively used to validate the quality of functional clothing during product development. They are a crucial part of user-oriented product development. The observed clothing prototypes are evaluated due to the wear trials, and after necessary modifications the final prototype is produced. A limited wear trial is performed in order to accept or reject a final solution (Rosenblad-Wallin, 1985).

The method for user-oriented product development presented by E. Rosenblad-Wallin has the following steps (Rosenblad-Wallin, 1985):

1. Identification of problem area
2. Problem analysis
3. Formulation of objective and project
4. Formulation of the user demands, the anticipated usage and general demands (based on user studies, interviews, environmental mappings, measurements, etc.)
5. Data processing and analysis
6. Specification of the use-demands and transformation of these into technical terms
7. Development of ideas and technical solution
8. Evaluation, modification and selection of prototype
9. Evaluation of the final solution in relation to the objectives

The sensation of comfort in clothing is usually tested during the wear trials with humans exposed to different environmental conditions, whether in controlled (climatic chambers) (Holmér, 1985) or uncontrolled (real environmental) conditions to obtain the interaction between the factors which influences the overall comfort sensation (Vokac et al., 1976; Li et al., 1988).

The controlled artificial environment involves specific protocol during the wear trials and the complete control of the environmental parameters such as the air temperature, the relative humidity, the air velocity, the duration of the exposures to a specific set of the environmental conditions and the activity taken by the subjects (Holmér, 1985).

The study by Meinander et al. compared the results of the thermal insulation properties of clothing systems defined through physical measurements using thermal manikins or through wear trials using human test subjects. The results showed that values from the manikin tests are reasonably reproducible; however, there were greater individual differences in the wear trial results. The thermal insulation values defined with thermal manikins correspond well with the wear trial values at moderate temperatures in the range of 273.15 to 283.15 K (0 to 10°C). However, at 298.15 K (25°C), a correction for sweating is in order (Meinander et al., 2004).

4.2 IMPROVING COMFORT IN TEXTILES AND CLOTHING AND FUTURE TRENDS

In the last decades, most of the testing methods of thermal comfort, protective performance and ergonomics of clothing focused on the straight standing posture. This is only one of the many postures during working conditions, and further investigation should be performed in order to classify potential shortcomings in both protective clothing designs and to address potential threats to the human body (Mert et al., 2017).

The various fibre types, fabric structures, textile constructions and finishes, novel garment engineering techniques, smart textiles and wearable electronics lay the basis for the sophisticated clothing system design. The performance requirements of many products demand the balance of widely different properties such as drape, thermal insulation, moisture management, antistatic and stretch properties and improved physiological comfort (Shishoo, 2015). Nowadays, a breakthrough in technology happens almost every single day thus enabling the production of a high variety of clothing with enhanced thermal and moisture-dissipating properties. From the phase-change materials to shape-memory alloys and all the way to electrically heating/cooling personal systems, these technology advancements work to prevent heat and cold stress in personal, functional, protective and sportswear.

The thermal and moisture properties of clothing can be enhanced in several ways (Havenith, 2005a; 2005b; Lee et al., 2007; McCann and Bryson, 2009; Mert et al., 2015a; Shishoo, 2015):

1. The appropriate choice of different textile materials; for example, by placing mesh inserts to increase air flow and moisture wicking in high sweat areas, using laminates (multilayer fabrics with a sandwich construction).
2. By using different pattern making techniques in order to get the optimum fit of garments in relation to the body shape or to design ventilation openings in order to increase the total heat loss. For example, a tight fit is necessary for physiological measurements systems and for body cooling. It is also important to remember that the thermal insulation decreases below a certain air layer thickness due to convection and the pumping effect inside the microclimate. When the air volume exceeds a certain limit, it becomes an ineffective insulator as forced convection due to the body movements prevents a linear increase of total dry heat loss.
3. The use of non-traditional bonding and joining techniques such as heat bonding, laser welding and moulding techniques to join mostly pieces of the technical materials from of synthetic fibres to provide the seamless fusing with waterproof properties.
4. The use of the layering system concept that incorporates an appropriate mix of innovative technical and other textile materials or assemblies in order to produce adequate evaporation and condensation effects. It can be accomplished by combining different layers of clothing garments or the combination of different material layers to address different needs within the clothing design.
5. By providing added value to the garment such as ease of movement, body support, energy storage and generation through new materials such as two-layer thermoplastic polyurethane film coating to store and release the body kinetic energy, antimicrobial, antiseptic and anti-UV protection by selecting fibres and finishes, embedding the wearable technology during the design process.

Functional clothing development accepted the concept of the layering system from military and sports apparel. Further development was driven by adopting lightweight waterproof textiles with waterproof coatings, which was fostered by the invention of synthetic fibres in the 1940s (McCann and Bryson, 2009).

The development of new materials quickens the improvements in thermal protective clothing and adjustment according to physical activity or environmental conditions (Wang and Gao, 2014).

Improved fibre spinning techniques helped produce fibres, yarns and fabrics with unique performance and high-tech characteristics (Shishoo, 2015).

The fibre spinning techniques and advances in polymer technology have improved substantially over the past few years and made it possible to produce more suitable fibres for use in protective clothing. Protective clothing performance requirements often demand the balance of widely different properties and place challenging demands on new technologies. Over the past few years, many new high-performance fibres with non-flammable properties have been modified and produced, such as m-aramid fibres. While the p-aramid fibres provide high strength and are mostly applicable for ballistic-protective clothing, the m-aramid fibres are used for heat and flame protective clothing (Shishoo, 2002).

The m-aramid fibres provide better flame and outstanding thermal protection used by brands like Nomex. Nomex® developed the aramid fibre with a high strength retention through extended heat exposure with good moisture regain. The polyphenylene sulphide (PPS), polyamide-imide (produced by Kermel), polyimide (produced by Lenzing P84), polybenzimidazole (produced by Hoechst-Celanese), novoloid or phenolic (produced by cured phenol-aldehyde fibres by Isover Saint Gobain), melamine-based fibres, chlorofibres polyvinyl chloride (produced under the industrial name of Rhovyl and Clevyl), polyacrylate (produced by index), semi-carbon (produced under the industrial name of Cellox and Panox) and polyetheretherketone (PEK) fibres also show excellent heat resistance, thermal stability and good comfort characteristics (Bajaj and Sengupta, 1992; Shishoo, 2000, 2002).

The novoloid fibres show high flame resistance; non-melting properties at any temperature, retention of textile integrity; no embrittlement or breakage, minimal evolution of smoke, nearly no shrinkage and virtually no toxic off-gassing during flame exposures. In general, they provide excellent thermal and electrical insulation (Shishoo, 2002).

Technical fibres such as viscose, polyester, nylon and acrylic/modacrylic fibres can be chemically modified during polymerisation. The surface modification for cellulose and wool fibres improves the heat resistance by adding fire-retardant finishes. The third example is using modifiers during the spinning process to gain flame retardation (Bajaj and Sengupta, 1992).

Nowadays, the highly functional fabrics are waterproof/moisture permeable and sweat-absorbing with high thermal insulation at low thickness values. Most of the materials and clothing development has always been influenced by textile innovation mostly as a direct result of military or aerospace research, such as Kevlar®, Teflon® and Gore-Tex® materials developed by NASA as protection against extreme temperatures for astronauts and commercially placed on the market by the DuPont from the 1970s (McCann and Bryson, 2009; Shishoo, 2015). Gore-Tex® is a microporous polymeric film made of polytetrafluoroethylene (PTFE). Coatings and membranes also produce improved textile properties. The hydrophilic and microfibrous membranes and microporous coatings are placed over the high functional fabrics to produce better water vapour transmission and permeability by modifying the porosity and thickness of the membranes and coatings laminated over the textile material (Shishoo, 2000).

The laminated or combined fabric is a material composed of two or more layers, bonded closely together by means of an added adhesive or by the adhesive properties of one or more of the component layers, at least one of which is a textile fabric and at least one of which is a substantially continuous polymeric layer (Fung, 2002). Laminated is considered to be different from a coated fabric, in that the layers are already pre-prepared and the second material can be a film, another fabric or some other material (Fung, 2002). Laminated textiles are made as fabric to fabric, fabric to foam, fabric to polymer or fabric to film bonding. The most famous type of laminated fabrics has a layer of membranes, which have micropores permeable to water vapour molecules but are impermeable to liquids and other organic molecules (Scott, 2005). So their primarily used to make textile waterproof, windproof and breathable, but they also protect against dirt, chemicals and pathogens. The types of membranes currently in usage are the hydrophilic solid membranes (polyurethane-PU, thermoplastic polyurethane-TPU, etc.) and hydrophobic microporous membrane (with micropores that are 100 times smaller than water droplets and have surface tension due to hydrophobic character, for example, PU and PTFE membranes).

Textile laminates with semi-permeable membranes are simultaneously permeable to water vapour from the inside, waterproof against moisture from the surroundings and impermeable to wind. However, since the membranes are thin, an additional insulating layer should be introduced to protect against low temperatures. They are also sensitive to punctures and abrasions and should be secured within the finished product. Clothing made of two-layer laminates should have additional material to secure the membrane mechanically and to reduce the negative coolness during contact with the skin. Laminates with a polytetrafluoroethylene (PTFE) membrane have minimal resistance to mass transport (minimal resistance to the permeability of water vapour), while laminates with polyurethane (PU) membrane ensures maximal resistance to mass transport through the laminate, which can cause discomfort (Sybilska and Korycki, 2010). The lamination techniques are flame lamination (to adhere polyurethane-PU foam to a textile material), adhesive lamination (to laminate two fabrics by applying an aqueous-based pressure-sensitive adhesive by knife-over-roller spreading), adhesive lamination-solvent-based (to laminate microporous membranes to textile fabrics to provide a barrier against liquids) and heat lamination (using a coating or hot-melt adhesive supplied as a solid, or slit film net or web can be carried out on the surface of a heated central drum) (Singha, 2012).

The coated fabric is the textile fabric on which there has been formed in situ, on one or both surfaces, a layer or layers of adherent coating material (Fung, 2002). The coating is a process in which a polymeric layer is applied directly to one or both surfaces of the fabric. Types of coating methods are direct coating (where the polymer resin compound is spread evenly over the surface of the fabric), foamed and crushed foam coating (where polymer is applied to woven, knitted fabrics and also to fabric produced from spun yarns), transfer coating (where the polymer is spread on to release paper to form a film and then to laminate this film to the fabric), hot-melt extrusion coating (thermoplastic polymers are applied as granules to the material and then melted between moving heated rollers), calender coating and rotary screen coating (Singha, 2012).

Traditional Lycra fibres allow high moisture transfer. The enhanced moisture management in hydrophilic laminates and microporous waterproof membranes, such as Gore-Tex®, resumed the path of increased material and functional clothing development and replaced the conventional PTFE membranes, which released the water vapour, however, not the droplets of sweat. Gore-Tex is a breathable waterproof fabric offering complete protection from water and wind, while simultaneously allowing body perspiration and heat release through a microporous membrane film laminated in between a lining and an outer shell textile. The X-BIONIC's Symbionic Membrane®, inspired by amphibian skin, is capable of transferring both water vapour and droplets of sweat from the microclimate between the skin and the clothing, thereby reducing the risk of heat stress. The Polartec® Power Stretch® wind resistant, abrasion resistant and breathable material keeps the skin dry using touch points on the fabric's inner surface to wick off the sweat to the outside of the fabric, where it spreads out to many times its original surface area, enabling it to dry at least two times faster than cotton (McCann and Bryson, 2009; Shishoo, 2015).

High-performance wicking materials are currently being investigated for us in next-to-skin clothing such as polyester fibres and nylon microfibers (Coolmax®, Thermostat®). They have the ability to transfer liquid moisture from the skin and dry quickly. With controlled release biocidal treatments, the hygiene properties are improved for during longer wear periods (Scott, 2000).

Except for textile-driven development, clothing technology has also based its development on increased health protection and heat/moisture management in combination with the application of smart textile innovations to address the anatomical, supporting and climate control demands of the body (McCann and Bryson, 2009).

Functional and smart textile development often adopts nature-based concepts, such as fir cones structures for creating a balance of breathability and waterproofness in phase-change textiles; the Nanosphere® nanoscale lotus leaf technology for moisture and stain repellent finishes, the second skin technology to enhance moisture wicking and ventilation; Polartec® fleece pile fabrics with variable patterning structures imitating animal fur to provide areas of added protection and ventilation and Patagonia BioMap® fleece fabrics with the variable-knit zones for warmth, dryness, cooling, mobility and fit (McCann and Bryson, 2009).

In recent years, smart textiles have started to be widely used to design clothing with added attributes. Smart textiles fall into the scope of a few categories. There are the passive smart textiles that can sense the environmental stimuli, the active smart textiles that can sense and react to the condition or stimuli; the very smart textiles with the ability to sense, react and adapt; and the intelligent textiles capable of responding in a pre-programmed manner. The passive smart properties include insulating, and moisture wicking structures, active smart include attributes such as antimicrobial protection, very smart include phase-change impact protection and thermal regulation, while the intelligent textiles permit power and signal pathways to be integrated into garments to monitor health functions and location (McCann and Bryson, 2009).

The new joining techniques are the third important factor that contributes to increased performance in the apparel industry. The hot air and sonic frictional heat bonding, ultrasonic and laser welding and moulding techniques are primarily applied to join pieces of the technical materials made from of synthetic fibres. They provide seamless fusing with waterproof properties. The laser cutting technology enabled clean edge cutting for the functional design details, such as waterproof zip openings, garment edges and perforations for ventilation. The seam-free knitting techniques are used in the structuring and shaping of intimate apparel, base-layer garments and medical hosiery. The laser finishes and textile bonding techniques also enable the encapsulation of miniaturised traditional and textile-based electronics; for example, the creation of the textile control buttons for electronic switches due to textile moulding (McCann and Bryson, 2009).

Protection against heat and absorption of UV radiation can be obtained due to ceramic fibres. The ceramic fibres are resilient-to-high temperatures, deformation and loss of tensile strength. They also pose low thermal conductivity, good insulating properties and have good chemical resistance (McCann and Bryson, 2009).

Today, the laminated fabric assemblies are widely used for protection against the extreme cold by trapping the still air between layers in order to increase thermal insulation and by preventing the wind chill effect. They also provide moisture withdrawal away from the skin to the outer layer, where it can more easily evaporate. The inner laminate layer promotes cooling by moisture dissipation (McCann and Bryson, 2009).

Nowadays, several smart textiles are already commercially available such as the phase-change materials (PCM), shape-memory materials (SMM), chromic materials (colour change) and conductive materials. The rapid progress in smart and other high-performance materials raised the interest of the sports and protective clothing industry (Lam and Stylios, 2006).

Dramatic temperature changes have been employed as the basis for PCM materials functioning. PCM is a thermal storage material used to regulate temperature fluctuations and the microcapsules are incorporated in textile structures to store energy when they change from solid to liquid and dissipate it when they change back from a liquid to solid (Mäkinen, 2006). The smart PCMs respond to changes in body temperature in absorbing excess heat and releasing it back to the body when required through the Matrix Infusion Coating technology (MIC), currently owned by Outlast®. The PCM interacts with the microclimate between the human body and the clothing and responds to the fluctuations in temperature by absorbing the heat when hot simultaneously creating a cooling effect. When the ambient temperature starts to decrease, the PCM materials release the absorbed heat creating a warming effect for the wearer (McCann and Bryson, 2009; Shishoo, 2015).

The most used PCMs in textiles are paraffin-waxes with various phase-change temperatures. The Outlast® adaptive textiles are based on PCM technology. The PCM-based textiles are currently used to design some outdoor clothing. However, to improve the thermal comfort characteristics of cold-protective clothing and the clothing ensemble in a cold environment by using PCM materials, they must produce enough heat in the garment layers to reduce heat loss from the body to the environment. The problem is that not all of the PCM microcapsules go through phase changes and consequently adequate thickness of the insulation system is still necessary for thermal comfort during extended exposure to cold environments. The usage of the PCM treated materials for flame resistant clothing also show certain limitations. When exposed to high environmental temperatures for short periods of time, PCMs may provide a short-term cooling effect, but the addition of paraffin-based PCMs may increase their flammability (Mccullough and Shim, 2006).

The ideal PCM for textile application should have a high heat of fusion, high heat capacity, high thermal conductivity, small volume change at the phase transition, be non-corrosive, non-toxic, non-flammable and exhibit little or no decomposition or supercooling (Bendkowska, 2006).

The desired thermal performance of the textile product due to PCM technology is obtained due to the selection of the appropriate PCM and choosing the sufficient quantity of the PCM applied to the appropriate fibrous substrate. Functions of the textile structure with PCM technology are defined as (Bendkowska, 2006):

1. A cooling effect (heat absorption by the PCM)
2. A heating effect (heat emission from the PCM)
3. A thermo-regulating effect that keeps the temperature of the surrounding substrate approximately constant (heat absorption or heat emission of the PCM)
4. An active thermal barrier effect that regulates the heat flux through the substrate and adapts the heat flux to thermal needs (heat absorption or heat emission of the PCM)

Three main categories of smart material can be distinguished (Boussu and Petitniot, 2006):

1. The piezoelectric materials that generate an electrical tension when submitted to a strain or inversely an electrical tension can provide a strain.
2. The magnetostrictive or electrostrictive materials that change their shape under a magnetic or respectively electrical field.
3. The shape-memory alloy (SMA) materials that modify their crystal lattice structure due to an external stimulus.

SMMs change their shape from a temporary deformed shape to a previously 'programmed' shape due to an external stimulus, such as changes in the surrounding temperature, by elastic energy, magnetic field, electric field, pH-value, UV light or water. The basic working principle of SMMs is dramatic deformations in a stress-free recovery due to specific recovery. The situation where SMA deforms under a load is called restrained recovery, where the SMMs is prevented from recovering from the initial strain, thus inducing the recovery tensile stress and the work can be performed by the SMMs. The two main phenomena of SMAs are the shape-memory effect (the shape change) or the superelasticity of the material (Honkala, 2006).

Shape-memory technology is nowadays more frequently applied in the textile and apparel industry. Shape-memory polymers (SMPs) and stimuli-sensitive polymers (SSPs) are polymeric materials with the ability of easy shaping, high shape stability and adjustable transition temperature, thus having a better potential for textile/clothing products. They can be used for smart material or functional material applications. The SSPs are used to enhance textile properties such as air permeability, hydrophilicity, heat transfer, shape and light reflectance, while the SMPs offer greater deformation capacities, easier shaping and greater shape stability, and small changes to the chemical structure and composition of SMPs result in a wide variety of transition temperatures and mechanical properties. Today, the prominent application of the shape-memory polymers film for breathable textiles in the sportswear due to the excellent heat retaining property at low temperature and gas permeability at high temperature. The smart fibres from SMP are used for medical applications such as sealing difficult wounds with limited access, for stents and screws for holding bones together, for surgical protective garments, bedding, diapers, training pants and feminine care and incontinence products. The SMPs films can be incorporated as an interliner in multilayer garments and used as adaptable thermal insulation in the outdoor protective clothing or leisurewear industry. The fabrics coated or laminated with the SMP form an ideal combination of thermal insulation and vapour permeability in the human body microclimate for underwear and outerwear clothing (Hu and Mondal, 2006a, 2006b).

Some shape-memory polymers are also simultaneously used as a sensor and/or actuator (Boussu and Petitniot, 2006).

SMAs, such as the nickel–titanium wires and polyurethane block copolymers films, automatically change shape and produce air gaps when the temperature changes. They ensure a constant air gap between fabric layers in intelligent heat or cold-protective clothing. These alloys return to a predetermined shape at a certain temperature after being deformed by an external force such as stretching, twisting or bending (Wang and Gao, 2014).

The specific possible properties of SMA (Boussu and Petitniot, 2006):

1. Superelasticity (the ability to buckle in a reversible process under stress)
2. Single memory effect (the ability to recover its initial shape under a thermal process after mechanical buckling)
3. Double memory effect (the to keep two different shapes at two different temperatures)
4. Rubbery effect (the ability of a residual buckling after the stress is removed; if the material is repeatedly subjected to stress the residual buckling will increase)
5. Damping effect (ability to absorb shocks or to reduce mechanical vibrations)

The future on application of the shape-memory materials for textile and apparel application will include (Hu and Mondal, 2006a):

1. Developing new shapes of polymer
2. Shaping recovery triggered by the room and body temperature
3. Studying the function and processing of the shape-memory polymers according to their structure and special properties
4. Understanding the complexity of phase transformation processes

Value of Globally Accepted Test Methods and Standards

5. Developing smart textiles with shape-memory technology
6. Establishing the test methods to evaluate the properties and performance of SMP and SMP based smart textiles

Cold-protective clothing is usually constructed by combining different material layers, such as by using the inter-dependent base-layer for sweat absorption through enhanced moisture wicking thus providing cooling by moisture dissipation and preventing the chilling effect, mid insulation layer for the thermal protection and outer protective layer, which is waterproof, windproof and breathable to allow the evaporation of sweat and heat, while simultaneously stopping the water entering the clothing (Wang and Gao, 2014), see Figure 4.3.

The thermal protection properties of clothing that serve to protect against the cold is dependent on the thermal resistance provided by textiles and overall thermal insulation provided by clothing. Other critical factors are the air permeability and the air trapped inside the textiles, and the transmission of water (Mathur et al., 1997). In cold-protective clothing, a fine balance between the optimum insulation and the thickness should always be considered. Higher values of basic clothing insulation usually correspond to thicker cold-protective clothing, which could limit movement and decrease physical work performance (Jussila et al., 2017). Thermal insulation is another critical factor when considering physiological reactions. Insufficient thermal insulation will lead to body cooling, while too high thermal insulation will lead to sweating and sweat accumulation (Jussila et al., 2010). Clothing for cold protection mostly tries to entrap as much air as possible by using the combination of three to four layers as air is one of the best insulators.

Since radiation forms a large part of this heat transfer in the cold, many textile researchers use low-emissivity metal coatings, fibrous batterings, and reflective

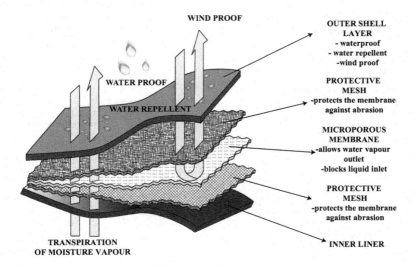

FIGURE 4.3 The heat and moisture exchange through materials layers for cold-protective clothing.

nano-fibrous thin layers to increase the thermal resistance of textiles (Chao et al., 2013; Morrissey and Rossi, 2015). A study by Morrissey and Rossi showed that the use of reflective insulation and interlayers, especially those with high air permeability and optical porosity insulation such as plasma-deposited metal coatings, can also increase the thermal resistance of cold-protective clothing (Morrissey and Rossi, 2015).

Clothing aimed at protecting against heat and flames is constructed in several layers as the clothing aimed to protect against the heat, with the major difference that the outer layer is aimed to protect against heat, flames and mechanical abrasions. The second layer in heat protective clothing serves as the moisture barrier to avoid sweat accumulation on the skin and is usually semi-permeable. The inner layer is thermal protective as opposed to cold-protective where the middle layer serves as thermal protection. Since these type of clothing serve to protect against highly hazardous environments with maximum possibility of burn injuries, the ventilation openings are avoided and heat and moisture transfer takes place in a transplanar direction enhancing the role of the material's qualities and layers and by reducing the thermal conductivity of these garments (Wang and Gao, 2014), see Figure 4.4.

Protective clothing used against hot liquid hazards involves different requirements in comparison to the cold and heat protective clothing. Blocking and minimising liquid penetration by the moisture barrier and decreasing liquid absorption in the fabric system are the most important requirements. Apart from these, the outer shell is usually impermeable with surface finishes decreasing mass transfer through the clothing system combined with the spacer material to optimise the thickness and protection properties, while the higher thermal insulation is necessary for reducing the heat transfer to human skin.

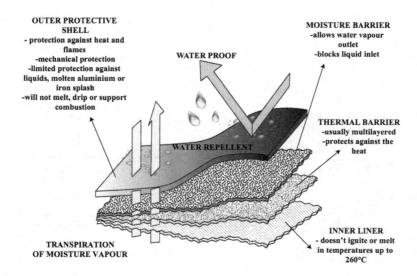

FIGURE 4.4 The heat and moisture exchange through materials layers in clothing aimed to protect against the heat and flames.

Chemical, biological, radiological and nuclear (CBRN) protective clothing should in comparison protect the respiratory tract, and the whole body, against vapour and liquid. These protective suits have the most all-around and complicated requirements, involving ease of movement, anti-gas leakage, liquid-tight integrity, permeation resistance to various chemicals, flammability resistance, an anti-burst property, anti-puncture and tear resistance, cut resistance, abrasion resistance and slip resistance, depending on the nature of the usage. With protective clothing used against hot liquid hazards and CBRN protective clothing, there is a real danger in heat stress, which is why this type of clothing should be cautiously designed. There is also a variety of physiological and psychological components that have to be covered, such as an increase in oxygen consumption, headaches, nausea, hunger, discomfort, body movement reduction, impaired dexterity and tactility, anxiety and claustrophobia, decreased and distorted field of vision, dehydration, decrease in cognitive performance, hyperventilation, claustrophobia, compulsive practices, obsessive actions, depression and even loneliness. Ballistic-protective clothing and body armour systems primarily serve as protection against conventional munitions and explosive devices. In addition, they also have to protect against temperature, humidity, solar radiation, wind, rain, hail, snow and ice, dust and sand, atmospheric pressure and electricity and biological hazards. However, they place a great deal of thermal burden on the wearer, since they are heavy, bulky and stiff, with low permeability thus preventing microclimate heat and vapour exchange. In order to decrease the thermal burden, reduction in the mass and number of layers should be made by a vast variety of personal cooling strategies such as liquid- or air-cooling systems, phase-change materials and extremity cooling systems. In addition, the thermal and moisture requirements are different for other types of protective clothing such as spacesuits and medical protective clothing. For instance, spacesuits have to protect against extreme cold or extreme heat and the vacuum of space during an extravehicular activity or the differences in air pressure during an intravehicular activity. Thermal protection and comfort are limited. The outer layer cannot enhance the zero gravity without modelling the composition of the hard outer shell covering the torso, using the protective fabrics and multiple layers of insulation. Spacesuits are designed to adjust to changes in thermal protective needs and be energy efficient. Medical gowns should be liquid-proof and pathogen-proof, which often results in the usage of the non-breathable coated materials without the pores. The pores would enhance the danger of liquid or bacteria penetration. However, this puts the wearer in thermal stress due to the sweat accumulation under surgical gowns. Other reasons for thermal discomfort are the air temperature requirements, since the air temperature within the operating rooms should be kept at 297.15 to 299.15 K (24–26°C) and high humidity levels (40–60%) to protect a patient from a risk of hypothermia. Second, some operations require the surgeons to stand and perform operations for a prolonged time in a standing position. Future development aims to enhance moisture vapour permeation through membranes, water-resistant coatings, flexible, elastic, lightweight and thin waterproof materials with high strength and resistance to tear and puncture damage (Wang and Gao, 2014).

Lately, an innovative light, flexible and active wearable thermal protection in the form of artificially intelligent technology has begun development, since the traditional passive thermally insulated clothing is insufficient, too blocky and heavy thus constraining the movement of wearers. This active thermal-functional clothing is an intelligent barrier with electric components (embedded electrodes, cables and wires) integrated into the fabric's pattern design. The modern wearable electric thermal technology integrates both textiles and electrical engineering to closely fit the body with free motions and flexibility and evenly heating with less weight burdened simultaneously. The future development of these active, intelligent wearable systems involves enabling the thermal-functional fabrics to collect physical/biological/chemical signals from the wearers and allow for reagent storage/manipulation, such as storing encapsulated medicine inside paired with the electric heating ability (Chenxiao and Li, 2018). The application of the personal heating garments (PHG) such as the electrical heating garments (EHG), the phase-change material garments, the chemical heating garments and the fluid/air flow heating garments, enabled a wider range of operating temperatures and improved its protection against the cold (Wang et al., 2010).

The humidity-induced, bendable smart clothing has been designed to reversibly adapt their thermal insulation functionality to the utilisation of humidity sensitive, smart polymer materials. There are two types of reversible humidity sensitive clothing, one designed to mimic the pores in human skin and the other works as adjustable thickness clothing. The first type of clothing has pre-cut pores in Nafion™ sheets (perfluorosulfonic acid polymer). When the humidity increases during human sweating, the pores permit the air flow. The flaps close automatically after perspiration to keep the wearer warm. The second type has the bent polymer sheets inserted between two fabrics, which become thinner when the humidity increases and reduces the gap between the two fabrics to reduce the thermal insulation or recover its thickness when the humidity reduces (Zhong et al., 2017).

REFERENCES

Advanced Characterization and Testing of Textiles. 2018. ed. P. Dolez, O. Vermeersch and V. Izquierdo, Cambridge, MA: Elsevier Ltd., ISBN 978-0-08-100453-1.

Ahn, H. W., C. H. Park and S. E. Chung. 2010. Waterproof and breathable properties of nanoweb applied clothing. *Textile Research Journal* 81(4): 1438–1447

ANSI/ASHRAE Standard 55-2017 Thermal Environmental Conditions for Human Occupancy. 2017. Atlanta, GA: American Society of Heating, Refrigerating and Air Conditioning Engineers Inc. (ASHRAE), ISSN 1041-2336.

ASTM D 1518-14 Standard Test Method for Thermal Resistance of Batting Systems Using a Hot Plate. 2014. Book of Standards Volume: 07.01. Philadelphia, PA: ASTM International.

ASTM E 96 / E 96M-16 Standard Test Methods for Water Vapor Transmission of Materials. 2016. Book of Standards Volume: 04.06. Philadelphia, PA: ASTM International, www.astm.org.

ASTM F 1291-16 Standard Test Method for Measuring the Thermal Insulation of Clothing Using a Heated Manikin. 2016. Book of Standards Volume: 11.03. Philadelphia, PA: ASTM International.

ASTM F 1868-02 Standard Test Method for Thermal and Evaporative Resistance of Clothing Materials Using a Sweating Hot Plate. 2002. Book of Standards Volume: 11.03. Philadelphia, PA: ASTM International.

ASTM F 1868-17 *Thermal and Evaporative Resistance of Clothing Materials Using a Sweating Hot Plate Test.* 2017. Book of Standards Volume: 11.03. Philadelphia, PA: ASTM International.

ASTM F 2298-03(2009)e1 *Standard test method for water vapour diffusion resistance of clothing materials using the dynamic moisture permeation cell.* 2009. Book of Standards Volume: 11.03. Philadelphia, PA: ASTM International (currently withdrawn).

ASTM F 2370-16 *Measuring the Evaporative Resistance of Clothing Using a Sweating Manikin.* 2016. Book of Standards Volume: 11.03. Philadelphia, PA: ASTM International.

ASTM F 2371-16 *Standard Test Method for Measuring the Heat Removal Rate of Personal Cooling Systems Using a Sweating Heated Manikin.* 2016. Book of Standards Volume: 11.03. Philadelphia, PA: ASTM International.

ASTM F 2732-16 *Standard Practice for Determining the Temperature Ratings of Cold Weather Clothing.* 2016. Book of Standards Volume: 11.03. Philadelphia, PA: ASTM International.

ASTM F 372-73 *Standard Test Method for Water Vapor Transmission Rate of Flexible Barrier Materials Using an Infrared Detection Technique.* 1976. Annual Book of ASTM Standards, Part 21, Vol. 04.06. Philadelphia, PA: ASTM International, 367–372.

Bajaj, P. and A. K. Sengupta. 1992. Protective clothing. *Textile Progress* 22(2–4), 1–110.

Bartels, V. T. and K. H. Umbach. 1999. Assessment of the physiological wear comfort of garments via a thermal manikin. In: *Proceedings of the 3rd International Meeting on Thermal Manikin Testing 3IMM at the National Institute for Working Life*, ed. H. O. Nilsson and I. Holmér. Stockholm, Sweden: National Institute for Working Life, ISBN: 91-7045-554-6.

Bedek, G., F. Salaün, Z. Martinkovska, E. Devaux and D. Dupont. 2011. Evaluation of thermal and moisture management properties on knitted fabrics and comparison with a physiological model in warm conditions. *Applied Ergonomics* 42: 792–800.

Bendkowska, W. 2006. Intelligent textiles with PCMs, Chapter 4. In: *Intelligent Textiles and Clothing.* Woodhead Publishing in Textiles, ed. H. R. Mattila, Cambridge, UK: Woodhead Publishing Ltd., ISBN 978-1-84569-005-2.

Bhatia, D. and U. Malhotra. 2016. Thermophysiological wear comfort of clothing: An overview. *Journal of Textile Science & Engineering* 6(2).

Bogdan, A. and M. Zwolińska. 2012. Future trends in the development of thermal manikins applied for the design of clothing thermal insulation. *Fibres & Textiles in Eastern Europe* 20(4–93): 89–95.

Boussu, F. and J. Petitniot. 2006. Development of shape memory alloy fabrics for composite structures, Chapter 8. In: *Intelligent Textiles and Clothing.* Woodhead Publishing in Textiles, ed. H. R. Mattila. Cambridge, UK: Woodhead Publishing Ltd., ISBN 978-1-84569-005-2.

BS 3424-12:1996. *Testing Coated Fabrics. Accelerated Ageing Tests.* 1996. BSI.

BS 7209:1990. *Specification for Water Vapour Permeable Apparel Fabrics.* 1990. BSI.

BS EN 12280-1:1998. *Rubber- or Plastics-Coated Fabrics. Accelerated Ageing Tests. Heat Ageing.* 1998. BSI.

BS EN 12280-3:2002. *Rubber- or Plastics-Coated Fabrics. Accelerated Ageing Tests. Environmental Ageing.* 2002. BSI.

Burton, A. C. and O. G. Edholm. 1955. *Man in a Cold Environment.* London, UK: Edward Arnold Ltd.

Chao, S., F. Jintu, W. Huijun, W. Yuenshing and W. Xianfu. 2013. Cold protective clothing with reflective nano-fibrous interlayers for improved comfort, *International Journal of Clothing Science and Technology* 25(5): 380–388.

Cheng, Y., J. Niu and N. Gao. 2012. Thermal comfort models: A review and numerical investigation. *Building and Environment* 47:13–22.

Chenxiao Y. and L. Li. 2018. The application wearable thermal textile technology in thermal-protection applications. *Trends in Textile & Fashion Design* 1(2): 1–9.

Chinta, S. K. and D. Satish. 2014. Studies in waterproof breathable textiles. *International Journal of Recent Development in Engineering and Technology* 3(2): 16–20.

Das, S. and V. K. Kothari. 2012. Moisture vapour transmission behaviour of cotton fabrics. *Indian Journal of Fibre & Textile Research* 37: 151–156.

de Dear, R. J., E. Arens, Z. Hui, M. Oguro. 1997. Convective and radiative heat transfer coefficients for individual human body segments. *International Journal of Biometeorology*, 40: 141–156.

DIN 53924:1997-03. *Prüfung von Textilien – Bestimmung der Sauggeschwindigkeit von textilen Flächengebilden gegenüber Wasser (Steighöhenverfahren)*. 1997. DIN.

Endrusick, T. L., L. A. Stroschein and R. R. Gonzalez. 2001. U.S. military use of thermal manikins in protective clothing research, Defense Technical Information Center Compilation Part Notice, ADP012410. In: *Blowing Hot and Cold: Protecting Against Climatic Extremes*, ADA403853. http://www.dtic.mil/dtic/tr/fulltext/u2/p012410.pdf (accessed on August 20, 2018).

Fan, J., X. Cheng, Y.-S. Chen. 2003. An experimental investigation of moisture absorption and condensation in fibrous insulations under low temperature. *Experimental Thermal and Fluid Science* 27: 723–729.

Farrington, R. B., J. P. Rugh, D. Bharathan and R. Burke. 2004. Use of a thermal manikin to evaluate human thermoregulatory responses in transient, non-uniform, thermal environments. *2004 SAE International*. https://www.nrel.gov/transportation/assets/pdfs/20 04_01_2345.pdf (accessed on April 15, 2018).

Fung, W. 2002. *Coated and Laminated Textiles*. Cambridge, UK: Woodhead Publishing Ltd, ISBN 1-85573-576-8.

Gagge, A. P., A. C. Burton and H. C. Bazzet. 1941. A practical system of units for the description of the heat exchange of man with his environment. *Science* 94: 428–430.

Gagge, A. P., J. A. J. Stolwijk and J. D. Hardy. 1967. Comfort and thermal sensations and associated physiological responses at various ambient temperatures. *Environmental Research* 1: 1–20.

Gao, C., I. Holmér, J. Fan, X. Wan, J. Y. S. Wu and G. Havenith. 2006. The comparison of thermal properties of protective clothing using dry and sweating manikins. In: *Proceedings of the 3rd European Conference on Protective Clothing (ECPC) and NOKOBETEF 8*. Gdynia, Poland: Central Institute for Labour Protection.

Gibson, P. 1994. *Governing Equations for Multiple Heat and Mass Transfer in Hygroscopic Porous Media with Applications to Clothing Materials*. Aberdeen Proving Ground, MD: U. S. Army Natick Research, Development and Engineering Center (NSRDEC), Technical Report, NATICK/TR-95/004.

Hardy, J. D. 1937. The physical laws of heat loss from the human body. *Proceedings of the National Academy of Science* 23: 631–637.

Havenith, G. 2005a. Thermal conditions measurement, Chapter 60. In: *Handbook of Human Factors and Ergonomics Methods*, ed. N. Stanton, A. Hedge, K. Brookhuis, E. Salas and H. Hendrics. Boca Raton, FL: CRC Press, 552–573, ISBN 0-415-28700-6.

Havenith, G. 2005b. Clothing heat exchange models for research and application. In: *Proceedings of 11th International Conference on Environmental Ergonomics*, ed. I. Holmér, K. Kuklane and C. Gao. Ystad, Sweden: Thermal Environment Laboratory, EAT, Department of Design Sciences, Lund University, 66–73, ISSN: 1650-9773.

Havenith, G. and R. Heus. 2000. Ergonomics of protective clothing. In: *Proceedings of Ergonomics of Protective Clothing Proceedings of NOKOBETEF 6 and 1st European Conference on Protective Clothing*, ed. K. Kuklane and I. Holmér. Stockholm, Sweden: National Institute for Working Life, 26–29, ISBN 91-7045-559-7.

Havenith, G., I. Holmér and K. Parsons. 2002. Personal factors in thermal comfort assessment: Clothing properties and metabolic heat production. *Energy and Buildings* 34: 581–591.

Havenith, G., R. Heus and W. A. Lotens. 1990. Resultant clothing insulation: A function of body movement, posture, wind, clothing fit and ensemble thickness. *Ergonomics* 33(1): 67–84.

Havenith. G. 2002. The interaction of clothing and thermoregulation. *Exogenous Dermatology* 1(5): 221–230.

Hes, L. 2004. The effective thermal resistance of fibrous layers in sleeping bags. *Research Journal of Textile and Apparel* 8(1): 14–19.

Hes, L. 2008. Non-destructive determination of comfort parameters during marketing of functional garments and clothing. *Indian Journal of Fibre & Textile Research* 33: 239–245.

Hes, L. and I. Dolezal. 1989. New method and equipment for measuring thermal properties of textiles, *Journal of Textile Machinery Society of Japan* 42(8): 71–75.

Hohenstein. Press release. Available at. https://www.hohenstein.de/en/inline/pressrelease_8 0768.xhtml (accessed on September 2, 2018).

Hollies, N. R. S. 1977. Psychological scaling in comfort assessment. In: *Clothing Comfort*, ed. N. R. S. Hollies and R. F. Goldman. Ann Arbor, MI: Ann Arbor Science Publishers Inc., 107–120.

Hollies, N. R. S., A. G. Custer, C. J. Morin and M. E. Howard. 1979. A human perception analysis approach to clothing comfort, *Textile Research Journal* 49(10): 557–564.

Holmér, I. 1985. Heat exchange and thermal insulation compared in woollen and nylon garments during wear trials. *Textile Research Journal* 55(9): 511–518.

Holmér, I. 1999. Thermal manikins in research and standards. In: *Proceedings of the Third International Meeting on Thermal Manikin Testing 3IMM at the National Institute for Working Life*, ed. H. O. Nilsson and I. Holmér. Stockholm, Sweden: National Institute for Working Life, ISBN: 91-7045-554-6.

Holmér, I., H. Nilsson, G. Havenith and K. Parsons. 1999. Clothing convective heat exchange proposal for improved prediction in standards and models. *Annals of Occupational Hygiene* 43(5): 329–337.

Holmes, D. A. 2000. Performance characteristics of waterproof breathable fabrics. *Journal of Industrial Textiles* 29(4): 306–316.

Honkala, M. 2006 Introduction to shape memory materials, Chapter 6. In: *Intelligent Textiles and Clothing*. Woodhead Publishing in Textiles, ed. H. R. Mattila, Cambridge, UK: Woodhead Publishing Ltd., ISBN 978-1-84569-005-2.

Hu, J. and S. Mondal. 2006a. Temperature-sensitive shape memory polymers for smart textile applications, Chapter 7. In: *Intelligent Textiles and Clothing*. Woodhead Publishing in Textiles, ed. H. R. Mattila. Cambridge, UK: Woodhead Publishing Ltd., ISBN 978-1-84569-005-2.

Hu, J. and S. Mondal. 2006b. Study of shape memory polymer films for breathable textiles, Chapter 9. In: *Intelligent Textiles and Clothing*. Woodhead Publishing in Textiles, ed. H. R. Mattila. Cambridge, UK: Woodhead Publishing Ltd., ISBN 978-1-84569-005-2.

Huang, J. 2006. Sweating guarded hot plate test method. *Polymer Testing* 25: 709–716.

Huang, J. 2007. A new test method for determining water vapor transport properties of polymer membranes. *Polymer Testing* 26: 685–691.

ISO 11092:2014 Textiles-Physiological Effects – Measurement of Thermal and Water-Vapour Resistance under Steady-State Conditions (Sweating Guarded-Hotplate Test). 2014. ISO-International Organization for Standardization.

ISO 12572:2016 Hygrothermal Performance of Building Materials and Products – Determination of Water Vapour Transmission Properties – Cup Method. 2016. ISO-International Organization for Standardization.

ISO 1419:2018 Rubber- or Plastics-Coated Fabrics – Accelerated-Ageing Tests. 2018. ISO-International Organization for Standardization..

ISO 15496:2018 Textiles – Measurement of Water Vapour Permeability of Textiles for the Purpose of Quality Control. 2018. ISO-International Organization for Standardization..

ISO 15831:2004 Clothing – Physiological Effects – Measurement of Thermal Insulation by Means of a Thermal Manikin. 2004. ISO-International Organization for Standardization.

ISO 2528:2017 Sheet Materials – Determination of Water Vapour Transmission Rate (WVTR) – Gravimetric (Dish) Method. 2017. ISO-International Organization for Standardization.

ISO 7730:2005 Ergonomics of the Thermal Environment – Analytical Determination and Interpretation of Thermal Comfort using Calculation of the PMV and PPD Indices and Local Thermal Comfort Criteria. 2005. ISO-International Organization for Standardization.

ISO 9920:2007 Ergonomics of the Thermal Environment – Estimation of Thermal Insulation and Water Vapour Resistance of a Clothing Ensemble. 2007. ISO-International Organization for Standardization.

Jürgens, W. 1924. *Der Warmeubergang an einer ebenen Wand: Arbeiten aus dem Heiz- und Lüftungsfach, Beihefte zum Gesundheits-Ingenieur, herausgegeben von der Schriftleitung des Gesundheits-Ingenieurs*, 19(1). München, Germany: R. Oldenbourg.

Jussila, K., A. Vaktskjold, J. Remes and H. Anttonen. 2010. The effect of cold protective clothing on comfort and perception of performance. *International Journal of Occupational Safety and Ergonomics (JOSE)* 16(2): 185–197.

Jussila, K., S. Rissanen, A. Aminoff, J. Wahlström, A. Vaktskjold, L.J. Talykova, J. Remes, S. Mänttäri and H. Rintamäki. 2017. Thermal comfort sustained by cold protective clothing in Arctic open-pit mining – A thermal manikin and questionnaire study. *Industrial Health* 55(6): 537–548.

Kaplan, S. and A. Okur. 2008. The meaning and importance of clothing comfort: A case study for Turkey. *Journal of Sensory Studies* 23: 688–706.

Kaplan, S. and S. Aslan. 2016. Subjective wear trials to evaluate thermal comfort of the foot clothing system including a sweat pad. *The Journal of The Textile Institute*. doi:10.1080/00405000.2016.1247680.

Kawabata, S. and M. Niwa. 1996. Recent progress in the objective measurement of fabric hand. In: *Macromolecular Concept and Strategy for Humanity in Science, Technology and Industry.* ed. Y. Ito, S. Okamura and B. Rånby. Berlin, Heidelberg: Springer-Verlag, ISBN: 978-3-642-64665-2.

Kothari, V. N. and K. Bal. 2005. Development of an instrument to study thermal resistance of fabrics. *Indian Journal of Fibre & Textile Research* 30: 357–362.

Kramar, L. 1989. Recent and future trends for high performance fabrics providing breathability and waterproofness. *Journal of Coated Fabrics* 28: 106–115.

Kurazumia, Y., T. Tsuchikawab, J. Ishiia, K. Fukagawa, Y. Yamatoc and N. Matsubara. 2008. Radiative and convective heat transfer coefficients of the human body in natural convection. *Building and Environment* 43: 2142–2153.

Lam Po Tang, S. and G. K. Stylios. 2006. An overview of smart technologies for clothing design and engineering. *International Journal of Clothing Science and Technology* 18(2): 108–128.

Lee, Y., K. Hong and S.-A. Hong. 2007. 3D quantification of microclimate volume in layered clothing for the prediction of clothing insulation. *Applied Ergonomics* 38: 349–355.

Li Y. 2001. The science of clothing comfort, *Textile Progress* 31(1–2): 1–135.

Li, Y., J. H. Keighley and I. F. G. Hampton. 1988. Physiological responses and psychological sensations in wearer trials with knitted sportswear. *Ergonomics* 31(11): 1709–1721.

Li, Y., J. H. Keighley, J. E. McIntyre and I. F. G. Hampton.1991. Predictability between objective physical factors of fabrics and subjective preference votes for derived garments. *Journal of Textile Institute* 82(3): 277–284.

Lotens, W. A. and G. Havenith. 1989. *Calculation of Clothing Insulation and Vapour Resistance*, TNO report. Netherlands: Netherlands organization for applied research.
Lotens, W. A. and G. Havenith. 1991. Calculation of clothing insulation and vapour resistance. *Ergonomics* 34(2): 233–254.
Mäkinen, M. 2006. Introduction to phase change materials, Chapter 3. In: *Intelligent Textiles and Clothing*. Woodhead Publishing in Textiles, ed. H. R. Mattila. Cambridge, UK: Woodhead Publishing Ltd., ISBN 978-1-84569-005-2.
Mathur, G. N., H. Raj and N. Kasturiya. 1997. Protective clothing for extreme cold. *Indian Journal of Fibre & Textile Research* 22: 292–296.
Mayor, T. S., S. Couto, A. Psikuta and R. M. Rossi. 2015. Advanced modelling of the transport phenomena across horizontal clothing microclimates with natural convection, *International Journal of Biometeorology* 59: 1875–1889.
McCullough, E. A. 2012. *Testing Services for the Evaluation of Fabric Systems, Clothing Systems, Personal Cooling Systems (PCS), and Sleeping Bag Systems*. Kansas State University, USA: Institute for Environmental Research.
McCullough, E. A. and H. Shim. 2006. The use of phase change materials in outdoor clothing, Chapter 5. In: *Intelligent Textiles and Clothing*. Woodhead Publishing in Textiles, ed. H. R. Mattila. Cambridge, UK: Woodhead Publishing Ltd., ISBN 978-1-84569-005-2.
McCullough, E. A., B. W. Jones and T. Tamura. 1989. A database for determining the evaporative resistance of clothing. *ASHRAE Transactions* 95(2): 316–328.
McCullough, E. and C. Kim. 1996. Insulation values for cold weather clothing under static and dynamic conditions. In: *Environmental Ergonomics Recent Progress and New Frontiers*, ed Y. Shapiro, D. Moran and Y. Epstein. London, UK: Freund Publishing House, 271–274.
Meinander, H., H. Anttonen, V. Bartels, I. Holmér, R. E. Reinertsen, K. Soltynski and S. Varieras. 2004. Manikin measurements versus wear trials of cold protective clothing (Subzero project). *European Journal of Applied Physiology* 92: 619–621.
Mert, E., A. Psikuta, M.-A. Bueno and R. M. Rossi. 2015a. Effect of heterogeneous and homogenous air gaps on dry heat loss through the garment. *International Journal of Biometeorology* 59: 1701–1710.
Mert, E., A. Psikuta, M.-A. Bueno and R. M Rossi. 2017. The effect of body postures on the distribution of air gap thickness and contact area. *International Journal of Biometeorology* 61: 363–375.
Mochida, T. 1977. Convective and radiative heat transfer coefficients for human body. *Bulletin of the Faculty of Engineering* 84: 1–11, Hokkaido University. http://hdl.handle.net/2115/41421 (accessed on August 20, 2018).
Morrissey, M. P. and R. M. Rossi. 2015. Recent developments in reflective cold protective clothing. *International Journal of Clothing Science and Technology* 27(1): 17–22.
Murakami, S., J. Zeng, T. Hayashi. 1999. CFD analysis of wind environment around human body. *Journal of Wind Engineering and Industrial Aerodynamics* 83(1–3): 393–408.
Nazaré, S. and D. Madrzykowski. 2015. A review of test methods for determining protective capabilities of fire fighter protective clothing from steam, *NIST Technical Note 1861*. National Institute of Standards and Technology (accessed on December 12, 2017).
Nicholson, G. P., J. T. Scales, R. P. Clark and M. L. de Calcina-Goff. 1999. A method for determining the heat transfer and water vapour permeability of patient support systems. *Medical Engineering & Physics* 21: 701–712.
Nielsen, R., B. W. Olesen and P. O. Fanger. 1985. Effect of physical activity and air velocity on the thermal insulation of clothing, *Ergonomics* 28(12): 1617–1631.
Nilsson, H. O. 2003. Comfort climate evaluation with thermal manikin methods and computer simulation models. *Indoor Air* 13(1): 28–37.
Nilsson, H. O. 2004. *Comfort Climate Evaluation with Thermal Manikin Methods and Computer Simulation Models*. Stockholm, Sweden: National Institute for Working Life, ISBN: 91-7045-703-4.

O'Brien, K., W. Koros and T. Barbari. 1986. A new technique for the measurement of multicomponent gas transport through polymeric films. *Journal of Membrane Science* 29: 229–238.

Oguro, M., E. Arens, R. J. de Dear, H. Zhang and T. Katayama. 2002. Convective heat transfer coefficients and clothing insulations for parts of the clothed human body under airflow conditions. *Journal of Architecture, Planning and Environmental Engineering* 561: 21–29.

Oliveira, A. V. M., V. J. Branco, A. R. Gaspar and D. A. Quintela. 2008. Measuring thermal insulation of clothing with different manikin control methods. Comparative analysis of the calculation methods. In: *Proceedings of the 7th International Thermal Manikin and Modelling Meeting*. Coimbra, Portugal: University of Coimbra. https://www.adai.pt/7i3m/Documentos_online/papers/11.oliveira_portugal.pdf (accessed on May 15, 2017).

OSH Wiki. ISO standards in the area of the ergonomics of the physical environment. Available at: https://oshwiki.eu/wiki/ISO_standards_in_the_area_of_the_Ergonomics_of_the_Physical_Environment (accessed on August 22, 2018).

Pamuk, O. 2008. Thermal manikins: an overview. *e-Journal of New World Sciences Academy* 3(1): 124–132.

Parsons, K. C. 2000a. Environmental ergonomics: A review of principles, methods and models. *Applied Ergonomics* 31: 581–594.

Parsons, K. C. 2000b. An adaptive approach to the assessment of risk for workers wearing protective clothing in hot environments. In: *Proceedings of Ergonomics of Protective Clothing Proceedings of NOKOBETEF 6 and 1st European Conference on Protective Clothing*, ed. K. Kuklane and I. Holmér. Stockholm, Sweden: National Institute for Working Life, 34–37, ISBN 91-7045-559-7.

Parsons, K. C., G. Havenith, I. Holmér, H. Nilsson, J. Malchaire. 1999. The effects of wind and human movement on the heat and vapour transfer properties of clothing. *Annals of Occupational Hygiene* 43(5): 347–352.

Pisacane, V. L., L. H. Kuznetz, J. S. Logan, J. B. Clark and E. H. Wissler. 2007. Use of thermoregulatory models to enhance space shuttle and space station operations and review of human thermoregulatory control. *Aviation, Space, and Environmental Medicine* 78(4): 48–55.

Protective Clothing: Managing Thermal Stress. 2014. Woodhead Publishing in Textiles: Number 154. ed. F. Wang and C. Gao. Cambridge, UK: Woodhead Publishing Ltd., ISBN 978-1-78242-032-3.

Pye, D., H. Hoehn and M. Panar. 1976. Measurement of gas permeability of polymers. II. Apparatus for determination of permeabilities of mixed gases and vapors, *Journal of Applied Polymer Science* 20: 287–301.

Rosenblad-Wallin, E. 1985. User-oriented product development applied to functional clothing design, *Applied Ergonomics*, 16(4): 279–287.

Scott, R. A. 2000. Fibres, textiles and materials for future military protective clothing. In: *Proceedings of Ergonomics of Protective Clothing Proceedings of NOKOBETEF 6 and 1st European Conference on Protective Clothing*, ed. K. Kuklane and I. Holmér, Stockholm, Sweden: National Institute for Working Life, 108–113, ISBN 91-7045-559-7.

Scott, R. A. 2005. *Textiles for Protection*, Cambridge, UK: Woodhead Publishing Ltd, ISBN 978-1-85573-921-5.

Sen, A. K. 2008. *Coated Textiles: Principles and Applications*, 2nd edition, Boca Raton, FL: CRC Press, ISBN 978-1-42005-345-6.

Shishoo, R. 2000. Innovations in fibres and textiles for protective clothing. In: *Proceedings of Ergonomics of Protective Clothing Proceedings of NOKOBETEF 6 and 1st European Conference on Protective Clothing*, ed. K. Kuklane and I. Holmér, Stockholm, Sweden: National Institute for Working Life, 79–87, ISBN 91-7045-559-7.

Shishoo, R. 2002. Recent developments in materials for use in protective clothing, *International Journal of Clothing Science and Technology*, 14(3):v201–215.

Singha, K. 2012. A review on coating & lamination in textiles: Processes and applications. *American Journal of Polymer Science* 2(3): 39–49.
Smart Clothes and Wearable Technology. 2009. Woodhead Publishing in Textiles: Number 83. ed. J. McCann and D. Bryson, Cambridge, UK: Woodhead Publishing Ltd., ISBN 978-1-84569-357-2.
Sybilska W. and R. Korycki. 2010. Analysis of coupled heat and water vapour transfer in textile laminates with a membrane. *Fibres & Textiles in Eastern Europe* 18(3,80): 65–69.
Takada, S., S. Hokoi and K. Nakazawa. 2004. Measurement of moisture conductivity of clothing, *Journal of the Human-Environmental System* 7(1): 29–34 (accessed on July 19, 2018).
Takada, S., S. Hokoi, N. Kawakami and M. Kudo. 1999. Effect of sweat accumulation in clothing on transient thermophysiological response of human body to environment. *Building simulation proceedings* 1: 385–392, http://www.ibpsa.org/proceedings/bs1999/bs99_c-15.pdf (accessed on July 19, 2018).
Textiles for Sportswear. 2015. Woodhead Publishing Series in Textiles: Number 162. ed. R. Shishoo, Cambridge, UK: Woodhead Publishing Ltd., ISBN 978-1-78242-229-7.
Umbach, K. H. 1988. Physiological tests and evaluation models for the optimisation of performance of protective clothing. In: *Environmental Ergonomics,* ed. I. B. Mekjavić, E. W. Banister and J. B. Morrison, London, UK: Taylor and Francis, 139–161.
Vogt, J. J., J.P. Meyer, V. Candas, J. P. Libert and J.C. Sagot. 1983. Pumping effects on thermal insulation of clothing worn by human subjects, *Ergonomics* 26(10): 963–974.
Vokac, Z., V. KØpke and P. Keül. 1972. Evaluation of the properties and clothing comfort of the scandinavian ski dress in wear trials, *Textile Research Journal* 42(2): 125–134.
Vokac, Z., V. KØpke and P. Keül. 1976. Physiological responses and thermal, humidity, and comfort sensations in wear trials with cotton and polypropylene vests, *Textile Research Journal* 46(1): 30–38.
Wang, F., C. Gao, K. Kuklane and I. Holmér. 2010. A review of technology of personal heating garments, *International Journal of Occupational Safety and Ergonomics (JOSE)* 16(3): 387–404.
Wang, G., W. Zhang, R. Postle and D. Phillips. 2003. Evaluating wool shirt comfort with wear trials and the forearm test, *Textile Research Journal* 73(2): 113–119.
Wang, Z., R. de Dear, M. Luo, B. Lin, Y. He, A. Ghahramani and Y. Zhu. 2018. Individual difference in thermal comfort: A literature review, *Building and Environment.* 138: 181–193. doi: 10.1016/j.buildenv.2018.04.040
Wehner, J., B. Miller and L. Rebenfeld. 1988. Dynamics of water vapor transmission through fabric barriers, *Textile Research Journal* 58(10): 581–592.
www.mercer-instruments.com/en/application/MVRT-WVRT-permaence-DVS.html (accessed on August 3, 2018).
Wyon. D. P. 1989. Use of thermal manikins in environmental ergonomics. *Scandinavian Journal of Work Environ Health* 15(1): 84–94.
Xua, X., P. Tikuisisb, R. Gonzaleza and G. Giesbrechtc. 2005. Thermoregulatory model for prediction of long-term cold exposure. *Computers in Biology and Medicine* 35: 287–298.
Yüksel, N. 2016. The review of some commonly used methods and techniques to measure the thermal conductivity of insulation materials, Chapter 6. In: *Insulation Materials in Context of Sustainability*, ed. A. Almusaed and A. Almssad, London, UK: Intech Open Ltd, 113–140, ISBN: 978-953-51-2625-6.
Zhang, C., X. Wang, Y. Lv, J. Ma and J. Huang. 2010. A new method for evaluating heat and water vapor transfer properties of porous polymeric materials, *Polymer Testing* 29: 553–557.

Zhong, Y., F. Zhang, M. Wang, C. J. Gardner, G. Kim, Y. Liu, J. Leng, S. Jin and R. Chen. 2017. Reversible humidity sensitive clothing for personal thermoregulation, *Scientific reports* 7:44208.

Zimmerli, T. 2000. Past, present and future trends in protective clothing. In: *Proceedings of Ergonomics of Protective Clothing Proceedings of NOKOBETEF 6 and 1st European Conference on Protective Clothing*, ed. K. Kuklane and I. Holmér, Stockholm, Sweden: National Institute for Working Life, 1–7, ISBN 91-7045-559-7.

5 Why Use Thermal Comfort Standards?

The standardisation of thermal comfort is of crucial significance since the thermal aspects of surrounding environments affect the physiological responses and working performances of humans, leading to more or less attained thermal comfort. Another reason to carefully address all of the factors influencing the thermal comfort state of humans is to avoid any potential health risks (Olesen et al., 2016).

The international standards, covering ergonomics of the thermal environment, also called the ergonomics of the physical environment (https://oshwiki.eu), including indoor and outdoor conditions and heat stress, cold stress and thermal comfort, represent the best available and agreed-upon information and methods used internationally to assess the impact of thermal conditions on health and comfort. As with all international standards, the standards in the field of the ergonomics of the thermal environment should be valid, reliable and useable with sufficient scope for practical application (Parsons, 2001; 2013). These international standards can be used in a complementary way and are divided up according to different environments (hot, moderate and cold). The remaining standards are divided into standards explaining human reactions to contact with solid surfaces and supporting standards (Parsons, 1999).

The general principles accepted by the international standards on thermal comfort involve all of the important factors affecting the human thermal comfort state. The goal is to standardise the evaluation methods and provide informative recommended limiting values for the different parameters or indices. These standards are then used to assess and design heating, ventilation and air conditioning systems (HVAC) and protective equipment to be used in moderate, cold and hot environments (Olesen, 1995). The main starting point considers the six main environmental and personal factors affecting the human response to thermal conditions. The second principle is based on the human response to local conditions and the quantification of these contributing factors. The third consideration of the international standards in the area of ergonomics of the thermal environment is the thermal strain imposed on the human body, leading to either heat or cold stress (Parsons, 2013).

Generally, international standards in the field of ergonomics have been divided according to different applications (Parsons, 1995a):

1. Basic standards on fundamental properties and human performance
2. Functional standards involving the characteristics of equipment, processes, products or systems
3. Environmental standards related to the effects of physical factors on human performance
4. Standards for test procedures and ergonomics data processing

The same distribution can be applied today. The requirements and the criteria for acceptable thermal climate conditions are described in standards such as EN ISO 7730 (ISO 7730:1994) and ASHRAE 55 (ANSI/ASHRAE Standard 55-2017). These criteria serve as guidance for general thermal comfort as based upon Fanger's thermal comfort equation. They are expressed through predicted mean vote (PMV) and predicted percentage of dissatisfied (PPD). They can also be expressed through a selected range of environmental parameters, such as operative temperature, air temperature and mean radiant temperature, air velocity and humidity. The local thermal discomfort is described by draught models, which observe the influence of the mean air velocity, turbulence intensity, air temperature, vertical air temperature differences, radiant temperature asymmetry and the surface temperature of the floor (Olesen, 2004). The acceptable conditions of thermal comfort defined by the standards are of major concern in determining energy use in buildings, the design approaches used to achieve those conditions, design and operation of all HVAC systems and feasibility of the cooling technologies (Fountain et al., 1999).

There is currently around 20 ISO standards in the field of the ergonomics of the thermal environment (Parsons, 2013; ISO 10551:1995; ISO 11079:2007; ISO 11092:2014; ISO 11399:1995; ISO 12894:2001; ISO 13732-1:2006; ISO 13732-2:2001; ISO 13732-3:2005; ISO 14415:2005; ISO 15265:2004; ISO 15743:2008; ISO 15831:2004; ISO 7243:2017; ISO 7726:1998; ISO 7730:1994; ISO 7730:2005; ISO 7933:2004; ISO 8996:2004; ISO 9886:2004; ISO 9920:2007), and around four ISO personal protective equipment for footwear standards (ISO 20344:2011; ISO 20345:2011; ISO 20346:2014; ISO 20347:2012).

Beside the International Organization for Standardization (ISO), the development of standards on thermal comfort and standards relating to the indoor environment is also carried out by the European Committee for Standardization (CEN) and American Society of Heating, Refrigerating and Air Conditioning Engineers (ASHRAE) (Olesen, 2004), which will be discussed in detail in Chapter 9.

ASTM (American Society for Testing and Materials) produced around seven existing standards in the scope of the test methods for assessing thermal and evaporative properties of fabrics and clothing (ASTM D1518-14, 2014; ASTM D1518-85, 2003; ASTM D7024-04, 2004; ASTM F1291-16, 2016; ASTM F1868-17, 2017; ASTM F2370-16, 2016; ASTM F2371-16, 2016). There is currently one ANSI/ASHRAE standard on thermal environmental conditions for human occupancy (ANSI/ASHRAE Standard 55-2017), and around five BS EN standards (British Standards) on safety, protective and occupational clothing/footwear for protection against cool, cold or heat (BS EN 14058:2004; BS EN 342:2004; BS EN 344-1:1993; BS EN 407:2004; BS EN 511:2006).

The standards are created to serve a greater good and to simplify everyday life, the design and production process of new products and to bring optimum quality requirements to both producers and users. Extensive research on the thermal comfort conditions carried out in the past (Fanger, 1970, 1973; Gagge, 1936; Winslow et al., 1937; Nevins et al., 1966; McNall et al., 1967; Mishra and Ramgopal, 2013) provided a deeper understanding of human thermal comfort preferences.

Standards play an important role in defining the thermal comfort preferences and defining the limits of heat stress for people exposed to severe environmental conditions, such as ISO 7243 (Epstein and Moran, 2006). Another example is the rising concern with risk assessment in the workplace. The main objective of risk assessment concerning the thermal working environment is not quantification, since correct quantification of the exposure and of the risk is very difficult and expensive (Malchaire, 2004). The strategy of risk management should rather work on prevention, reduction or even complete elimination of risk (Malchaire et al., 1999; Malchaire, 2006). The work to restructure legislation concerning the risk assessment at workplaces led to the creation of the international standard (ISO 15265:2004). Standards provide guidance on environmental design and control; they standardise the testing procedures and methods and also allow comparison, and consequently contribute to assessment and evaluation. Another important factor is health preservation due to defined conditions for thermal comfort (Parsons, 1995b).

Although in many countries users are involved through various working groups in standardising activities on a national level, during the early phases of standard development a greater need for user involvement and activation has been raised. However, the user is not easily defined. Ergonomics defines users usually as workers, engineers, equipment manufacturers, health and safety officers, ergonomists, etc. In other words, the users are considered as a beneficiary of the standardised method or practice. In the area of the ergonomics of the physical environment, most standards are developed by subject experts and rarely by groups of manufacturers and consumers. As Delleman et al. pointed out, the opinions of user groups will provide important perspectives but will not necessarily aid the creation of a useable standard, so usability testing must become part of the standardisation process (Delleman et al., 2000).

5.1 THE BASIC PRINCIPLES OF STANDARD APPROVAL

There is a basic difference between standards and regulations. Regulations specify legally enforceable requirements, non-compliance with which may be subject to sanctions, while standards are voluntary codes for which there are no legal obligations to comply. The content and structure of regulations are frequently subject to public consultation, but they are not necessarily the result of a consensual process, and everyone has an obligation to comply. The standards are often used as a means of demonstrating compliance with regulation, e.g. as with the 'new approach' directives of the EU. While compliance with standards is voluntary, in the case of litigation failure to comply with an existing standard might be deemed as demonstrating negligence (Hatto, 2001). Regulations are technical rules issued by government departments and agencies that provide clarity and guidance in their respective areas of responsibility. In the 1980s, a new approach was introduced in the EU which involved the delegation of the regulatory functions to private standardisation bodies. EU legislation is limited to laying down directives with mandatory essential requirements for health, safety, environmental and consumer protection, which cover the entire sector rather than a single product (Büthe and Mattli, 2011). On the other hand, standards as voluntary documents, in general, cover subjects like interoperability, interconnectability, levels of safety, performance, conformity, material properties, classification systems, testing methods, the operation of equipment or systems, quality assurance and definition of terms (Mattli, 2001).

Standards are detailed, technical documents that provide rules, guidelines or characteristics for a product or process, but they are not the law (regulation). For example, regulations typically set out only a general framework, procedure and/or set of standards to guard against a hazard, while standards are usually consistent with the law but are usually more complex and provide more detail. The standardising organisations are non-governmental organisations that form committees and try to reach a consensus on certain matters. They cannot force the law or employers to follow their standards. All they can do is make recommendations. Even though standards are voluntary and are not legally required, they represent a consensus on what experts consider as optimum and safe.

The voluntary, consensus-based standards play a large role in national and international infrastructures, economies and trade by providing agreed ways of naming, describing and specifying, measuring and testing, managing and reporting. In this way, the standards provide basic support for commercialisation, markets and market development; a recognised means for assuring quality, safety, interoperability and reliability of products, processes and services; a technical basis for procurement; a technical support for appropriate regulation and can lead to variety and cost reduction through optimisation and best practice. While the majority of standards address technical issues, such as the composition, treatment and testing of materials for different applications, there has been an increasing recognition over the last few decades that voluntary, consensus-based standards can contribute far more to business, and society in general, than simply technical specifications, testing methods and measurement protocols (Hatto, 2001).

The technical committee ISO TC 159 Ergonomics is the main organising structure within the ISO organisation, which controls and supervises the production of ISO standards within the scope of ergonomics applied to thermal comfort. Under the supervision of the technical committee TC 159, there are four subcommittees, which manage the new work projects relevant to their field of application. Finally, the work is further allocated to working groups (Parsons, 1995a), Figure 5.1.

The stages of the international standard production on ergonomics of the thermal environment (Parsons, 1995a, 2001):

1. The proposal is considered by the subcommittee and submitted for an international vote to the technical committee (TC), subcommittee (SC), working group (WG) and national standardising bodies (NSB).
2. If the proposal is accepted by the subcommittee ISO/TC 159 SC5 Ergonomics of the Physical Environment, the SC appoints the working group (ISO/TC 159 SC5 WG1 Ergonomics of the Thermal Environment) to write the working document and find the expert consensus.
3. Finding the consensus among all involved levels of hierarchical organisation in different countries (ISO/TC 159 Ergonomics, ISO/TC 159 SC5 Ergonomics of the Physical Environment, ISO/TC 159 SC5 WG1 Ergonomics of the Thermal Environment).
4. The international standard draft (Committee Draft) is assembled by the appointed working group and is distributed to be voted on and commented upon by member countries (around five months is allowed for voting).

Why Use Thermal Comfort Standards?

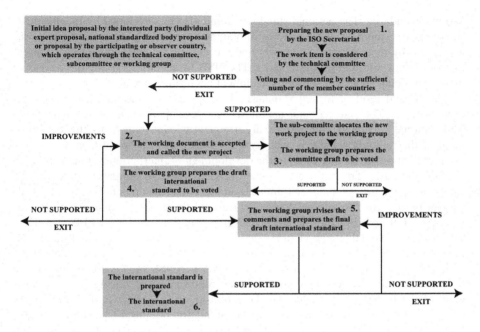

FIGURE 5.1 International standard preparation.

5. The preparation of the final draft of the proposed International Standard is performed by the working group and again distributed to be voted on as YES/NO among member countries.
6. The preparation of the International Standard is performed by the Secretariat of the ISO organisation.

The beginning of standard development starts based on the specific research activities. This research is undertaken prior to standardisation (normalisation) and called pre-normative research (PNR) and co-normative research (CNR). Pre-normative research involves all research steps necessary for demonstrating the feasibility, reliability and limitation of the technique or process to be standardised. Once the technique or process has been explored, a pre-standard, such as a publicly available specification (PAS) or technical specification (TS), is prepared and evaluated by potential users in a relatively short time. The availability of a pre-standard provides a basis for further research undertaken in conjunction with the standardisation process to establish a statistical basis for the technique or process, reproducibility, repeatability and uncertainty, and is called co-normative research (Hatto, 2010).

Standardisation is defined as an activity of establishing provisions for common and repeated use. The goal of standardisation is the achievement of the optimum degree of order in a given context with regard to actual or potential problems (Hatto, 2010).

The first level of formal standards development is the national standards bodies. These are government-recognised but not necessarily approved bodies, which publish national standards. The national standards could also be developed by independent standards development organisations, although this normally requires them to be accredited by their national standards body to ensure the development procedures satisfy certain minimum requirements. The NSB frequently facilitate the development of national standards through a technical committee structure.

The second level of formal standards development are international standards organisations. The international standards are developed on a consensual basis. The national technical committees are frequently contributing to the development of the regional standards such as CEN standards (European Committee for Standardization), and more often to the development of international standards such as ISO standards. The national members of CEN have an obligation to adopt all European standards as national standards and to withdraw any conflicting national standards that are in their catalogue. In Europe, there is the Vienna Agreement between CEN and ISO and the Dresden Agreement between CENELEC and IEC, which seek to avoid duplication and under which a number of international standards are adopted as European standards, usually without change (Hatto, 2001).

Each national member body (DIN, BSI, HZN, ANSI, etc.) sends a delegation of experts to the European committees as representatives during the development of the European standards. At the proposal stage, a proposal for a standard can come from a member of the European standards organisations, the European Commission or another European or international organisation. The European standards are developed by CEN, CENELEC (European Committee for Electrotechnical Standardization) and ETSI (European Telecommunications Standards Institute), formed of national bodies.

A majority, or 71% of the weighted majority among all NSB votes, is needed for the proposal to be accepted at the international level by the ISO. In addition, a sufficient number of the NSB must agree to participate. If there is an existing international standard on the subject, it will be adopted, unchanged, as the European standard. If this is not the case, the responsible working body will draw up a manuscript for the draft standard (prEN). The draft standard is distributed to the national standards organisations for commenting within three months. On the basis of the comments, the responsible working group can either publish the standard or issue a final draft. In a formal vote over a two-month period, the members then decide whether to accept this final draft as a European standard. Approval of the final draft requires at least 71% of the weighted votes of CEN members (www.din.de).

The difference between formal and informal standards is in the representation of those involved in the development and approval processes. The approval of informal standards is on an organisation and/or individual membership basis, while the formal standards are approved through national representation. Informal standard development follows similar procedures. The main difference is that the development and approval are undertaken by the members of the standards development organisations. They act as either individuals or representatives of their company or other organisation, rather than through a nationally constituted membership structure (Hatto, 2001). In both cases, the development and approval processes are based on the principle of consensus and general agreement (Hatto, 2001). Every standard is reviewed every 5 years (Parsons, 2001).

5.2 THE TYPES OF STANDARDS

The first general classification of standards categorises standards as formal, informal and private standards (Hatto, 2010). The formal standards are standards that are approved or adopted by one of the national, regional or international standards body. The informal standards are published by other standards development organisations, such as the American Society for Testing Materials International (ASTM International), the Institute of Electrical and Electronic Engineers (IEEE), etc. The companies develop private standards for internal use (Hatto, 2001, 2010).

The standards are further divided into normative or informative. The normative standards are those documents that contain requirements, which must be met in order for claims of compliance with the standard to be certified. The informative standards do not contain any requirements, and it is therefore not possible for compliance claims to be certified. While the majority of standards are normative, they typically also contain informative elements, in the form of notes, examples and informative annexes. The informative standards are typically published as technical reports and are not subject to periodic review but can be withdrawn if they are considered no longer relevant. There are also significant numbers of full international and European standards, primarily in the form of 'guidelines', which do not contain requirements. The normative documents are published as either technical specifications or full standards, usually containing the measurement and test methods, specifications or vocabularies (Hatto, 2001).

The standards can be classified as (APEC, 2010):

1. Basic standards standardise terminology, symbols and units. Among these, detailed standards including definitions and explanations of terminology are referred to as terminology standards and are sometimes organised separately.
2. Testing standards standardise tests, analyses, inspections and measurement methods. These standards solely prescribe measurement methods.
3. Inspection testing standards are sometimes created wherein various measurement conditions are established, and judgements are made using a pass/fail system for each measurement. These standards possess the characteristics of both testing standards and product standards.
4. Product standards standardise the shape, size material, components, quality, performance, durability, safety and function of products.
5. Process standards standardise the process such as the manufacturing procedure of the product. In particular, standards that standardise quality control systems and management are referred to as management standards.
6. Management standards aim at the continuous improvement of the management of the organisation.

The international standards for assessment of protective clothing are categorised as performance standards, which assess the protective function of protective clothing and evaluation standards, which evaluates the thermal stress on humans (Holmér, 1999).

Performance standards characteristics and usage (Holmér, 1999):

1. Determine the properties of clothing or materials.
2. Measure protection against external physical or chemical agents.
3. Are based on ambience-to-skin transfer.
4. Are based on defined environmental conditions.
5. Are based on material rather than clothing tests.

Evaluation standards (evaluation of physiological strain) characteristics and usage (Holmér, 1999):

1. Evaluate effects on the human body (strain).
2. Quantify heat transfer at the body surface.
3. Provide criteria for assessment of strain.
4. Apply to any environmental condition.
5. Apply to any activity level.
6. Require clothing ensemble values for critical properties.

5.3   THE BENEFITS PROVIDED BY STANDARDS

At a fundamental level, the objectives of standardisation are the dissemination of scientific and technical outcomes and the extension of community benefits through mutual understanding within society and assurance of public order. At a higher level, however, the objectives and significance of standardisation are manifold. Even a single standardisation project usually encompasses multiple objectives. There are many secondary benefits of standardisation, two of which are efficiency gains and economic advancement. The standardisation is expected to deliver benefits through (1) faster information communications and greater precision, (2) development of understanding among retailers and consumers at an earlier stage, (3) mitigation of conflicts between manufacturers and consumers, (4) enhancement of maintenance and repair efficiencies and (5) effective prevention of the recurrence of problems or accidents. When standardisation costs outweigh the standardisation benefits, however, the result is neither efficiency gains nor economic advancement. In such a case, standardisation should be avoided (APEC, 2010).

Basic functions of standards (APEC, 2010; Hatto, 2010):

1. Controlling diversity through simplification or in other words variety reduction/optimisation based on best practice.
2. Ensuring interoperability.
3. Ensuring compatibility.
4. Assuring quality and ensuring safety.
5. Disseminate information and reproducibility of measurement (standards on measurement and testing methods promote uniqueness for describing, quantifying and evaluating product attributes such as materials, processes and functions).

6. Ensuring corporate interests through:
   - Cheaper production and cost reductions (cost reductions through in-house standardisation, cost reductions through industry-consensus standardisation, cost reductions using standardisation and certification, assessing the cost reduction benefits).
   - Volume selling (creating, expanding, and sustaining markets, market creation and expansion through growth, market expansion by connecting to new markets, long-term market expansion).
   - Selling at higher prices (product differentiation, differentiation using certification, differentiation using testing standards).
7. Achieving policy goals.
8. Eliminating trade barriers.
9. Stimulating innovation.

Role of standards is to (Hatto, 2010):

1. Ensure the safety, quality and reliability of products, processes and services.
2. Ensure efficient production.
3. Ensure cost reduction through competition.
4. Support regulation.
5. Promote innovation and commercialisation through the dissemination of new ideas and good practice, validation of new measurement tools and methods and the implementation of new processes and procedures.

All European standards respond to the needs of the European Single Market. The European Single Market is based on the principle of free trade without barriers and with the simplest possible trading conditions. The legal requirements and standards within the single market aim at ensuring the free movement of goods while ensuring a high level of health and safety and other relevant protection for citizens. There are European standards for products, testing methods, business processes (such as procurement) and increasingly, for services. It is the policy of the European Commission, supported by the European Committee for Standardization (CEN) and the European Committee for Electrotechnical Standardization (CENELEC) and their members, that there should be a primacy of international standards in Europe. In other words, the ESOs do not develop European standards unless there is a specific European need. Instead, CEN, CENELEC and their members including BSI prefer to work at an international – as opposed to European – level (through ISO and IEC) and then adopt these standards in Europe as European Norms. The most common reason to develop a European or national standard in the absence of an international standard is to support a European or national regulatory requirement. Standards are also used in national markets as a way to deliver national government policies (*European Standards and the UK*).

Although much is already discovered on thermal comfort and human response to heat and cold, there are still human casualties due to the harsh environmental condition. This can be avoided by applying new technologies, techniques and indicators of thermal stain. There is also considerable work to be done in applying new work procedures, anticipating the influence of the forthcoming climate change, promoting the globalisation of standardised practices and developing standards for more diverse populations. All of these aspects provide significant challenges and require the continual review of existing standards (Parsons, 1999).

REFERENCES

ANSI/ASHRAE Standard 55-2017 Thermal Environmental Conditions for Human Occupancy. 2017. Atlanta, GA: American Society of Heating, Refrigerating and Air Conditioning Engineers Inc. (ASHRAE), ISSN 1041-2336.

ASTM D 1518-14 Standard Test Method for Thermal Resistance of Batting Systems Using a Hot Plate. 2014. Book of Standards Volume: 07.01. Philadelphia, PA: ASTM International.

ASTM D1518-85 Standard Test Method for Thermal Transmittance of Textile Materials. 2003. Philadelphia, PA: ASTM International.

ASTM D7024-04 Standard Test Method for Steady State and Dynamic Thermal Performance of Textile Materials. 2004. Philadelphia, PA: ASTM International.

ASTM F 1291-16 Standard Test Method for Measuring the Thermal Insulation of Clothing Using a Heated Manikin. 2016. Book of Standards Volume: 11.03. Philadelphia, PA: ASTM International.

ASTM F 1868-17 Thermal and Evaporative Resistance of Clothing Materials Using a Sweating Hot Plate Test. 2017. Book of Standards Volume: 11.03. Philadelphia, PA: ASTM International.

ASTM F 2370-16 Measuring the Evaporative Resistance of Clothing Using a Sweating Manikin. 2016. Book of Standards Volume: 11.03. Philadelphia, PA: ASTM International.

ASTM F 2371-16 Standard Test Method for Measuring the Heat Removal Rate of Personal Cooling Systems Using a Sweating Heated Manikin. 2016. Book of Standards Volume: 11.03. Philadelphia, PA: ASTM International.

BS EN 14058:2004 Protective Clothing – Garment for Protection Against Cool Environments. 2004. BSI.

BS EN 342:2004 Protective Clothing – Ensembles and Garment for Protection Against Cold. 2004. BSI.

BS EN 344-1:1993 Safety, Protective and Occupational Footwear for Professional Use. Requirements and Test Methods. 1993. BSI.

BS EN 407:2004 Protective Gloves Against Thermal Risks (Heat and/or Fire). 2004. BSI.

BS EN 511:2006 Protective Clothing – Protective Gloves Against Cold. 2006. BSI.

Büthe, T. and W. Mattli. 2011. The rise of private regulation in the world economy, Chapter 1. In: *The New Global Rulers: The Privatization of Regulation in the World Economy*. Princeton, NJ: Princeton University Press, ISBN 978-0-691-14479-5.

Delleman, N., T. Itani, K. Parsons, K. Scheuermann and T. Stewart. 2000. User involvement in international standardization. *Proceedings of the Human Factors and Ergonomics Society Annual Meeting* 44(35–6): 446–448.

Epstein Y. and D. S. Moran. 2006. Thermal comfort and the heat stress indices. *Industrial Health* 44: 388–398.

European Standards and the UK. London, UK: BSI Group. https://www.bsigroup.com/LocalFiles/en-GB/EUREF.pdf (accessed on April 10, 2017).

Fanger, P. O. 1970. *Thermal Comfort: Analysis and Applications in Environmental Engineering*. Copenhagen, Denmark: Danish Technical Press.

Fanger, P. O. 1973. Assessment of man's thermal comfort in practice. *British Journal of Industrial Medicine* 30: 313–324.

Fountain, M. E., E. A. Arens, T. Xu, F. S. Bauman and M. Oguru. 1999. An investigation of thermal comfort at high humidities. *ASHRAE Transactions* 105(2): 94–103

Gagge, A. P. 1936. The linearity criterion as applied to partitional calorimetry. *American Journal of Physiology* 31: 656–668.

Hatto. P. 2001. *Standards and Standardisation, A Practical Guide for Researchers.* European Commission, Directorate-General for Research & Innovation, Directorate G – Industrial technologies. https://ec.europa.eu/research/industrial_technologies/pdf/practical-standardisation-guide-for-researchers_en.pdf (accessed August 18, 2018).

Hatto. P. 2010. *Standards and Standardization Handbook.* Brussels, EU: European Commission, Directorate-General for Research & Innovation, Directorate G – Industrial Technologies. http://www.iec.ch/about/globalreach/academia/pdf/academia_governments/handbook-standardisation_en.pdf (accessed August 19, 2018).

Holmér, I. 1999. The role of performance tests, manikins and test houses in defining clothing characteristics relevant to risk assessment. *Annals of Occupational Hygiene* 43(5): 353–356.

ISO 10551:1995 Ergonomics of the Thermal Environment – Assessment of the Influence of the Thermal Environment Using Subjective Judgment Scales. 1995. ISO-International Organization for Standardization (also BS EN ISO 10551:2001).

ISO 11079:2007 Ergonomics of the Thermal Environment – Determination and Interpretation of Cold Stress When Using Required Clothing Insulation (IREQ) and Local Cooling Effects. 2007. ISO-International Organization for Standardization.

ISO 11092:2014 Physiological Effects – Measurement of Thermal and Water-Vapour Resistance under Steady-State Conditions (Sweating Guarded-Hotplate Test). 2014. ISO-International Organization for Standardization.

ISO 11399:1995 Ergonomics of the Thermal Environment – Principles and Application of Relevant International Standards. 1995. ISO-International Organization for Standardization.

ISO 12894:2001 Ergonomics of the Thermal Environment – Medical Supervision of Individuals Exposed to Extreme Hot or Cold Environments. 2001. ISO-International Organization for Standardization.

ISO 13732-1:2006 Ergonomics of the Thermal Environment – Methods for the Assessment of Human Responses to Contact with Surfaces – Part 1: Hot Surfaces. 2004. ISO-International Organization for Standardization.

ISO 13732-2:2001 Ergonomics of the Thermal Environment – Methods for the Assessment of Human Responses to Contact with Surfaces – Part 2: Human Contact with Surfaces at Moderate Temperature. 2001. ISO-International Organization for Standardization.

ISO 13732-3:2005 Ergonomics of the Thermal Environment – Methods for the Assessment of Human Responses to Contact with Surfaces – Part 3: Cold Surfaces. 2005. ISO-International Organization for Standardization.

ISO 14415:2005 Ergonomics of the Thermal Environment – Application of International Standards to People with Special Requirements. 2005. ISO-International Organization for Standardization.

ISO 15265:2004 Ergonomics of the Thermal Environment – Risk Assessment Strategy for the Prevention of Stress or Discomfort in Thermal Working Conditions. 2004. ISO-International Organization for Standardization.

ISO 15743:2008 Ergonomics of the Thermal Environment – Cold Workplaces – Risk Assessment and Management. 2008. ISO-International Organization for Standardization.

ISO 15831:2004 Clothing – Physiological Effects – Measurement of Thermal Insulation by Means of a Thermal Manikin. 2004. ISO-International Organization for Standardization.

ISO 20344:2011 Personal Protective Equipment – Test Methods for Footwear. 2011. ISO-International Organization for Standardization.

ISO 20345:2011 Personal Protective Equipment – Safety Footwear. 2011. ISO-International Organization for Standardization.

ISO 20346:2014 Personal Protective Equipment – Protective Footwear. 2014. ISO-International Organization for Standardization.
ISO 20347:2012 Personal Protective Equipment – Occupational Footwear. 2012. ISO-International Organization for Standardization.
ISO 7243:2017 Ergonomics of the Thermal Environment – Assessment of Heat Stress Using the WBGT (Wet Bulb Globe Temperature) Index. 2017. ISO-International Organization for Standardization (BS EN 27243:2017).
ISO 7726:1998 Ergonomics of the Thermal Environment – Instruments for Measuring Physical Quantities. 1998. ISO-International Organization for Standardization.
ISO 7730:1994 Ergonomics of the Thermal Environment – Analytical Determination and Interpretation of Thermal Comfort Using Calculation of the PMV and PPD Indices and Local Thermal Comfort Criteria. 1994. ISO-International Organization for Standardization.
ISO 7730:2005 Ergonomics of the Thermal Environment – Analytical Determination and Interpretation of Thermal Comfort Using Calculation of the PMV and PPD Indices and Local Thermal Comfort Criteria. 2005. ISO-International Organization for Standardization.
ISO 7933:2004 Ergonomics of the Thermal Environment – Analytical Determination and Interpretation of Heat Stress Using Calculation of Predicted Heat Strain. 2004. ISO-International Organization for Standardization.
ISO 8996:2004 Ergonomics of the Thermal Environment – Determination of Metabolic Rate. 2004. ISO-International Organization for Standardization.
ISO 9886:2004 Ergonomics – Evaluation of Thermal Strain by Physiological Measurements. 2004. ISO-International Organization for Standardization.
ISO 9920:2007 Ergonomics of the Thermal Environment – Estimation of Thermal Insulation and Water-Vapour Resistance of a Clothing Ensemble. 2007. ISO-International Organization for Standardization.
Malchaire, J. B. 2004. The SOBANE risk management strategy and the Deparis method for the participatory screening of the risks. *International Archives of Occupational and Environmental Health* 77: 443–450.
Malchaire, J. B. 2006. Participative management strategy for occupational health, safety and well-being risks. *Giornale italiano di medicina del lavoro ed ergonomia* 28(4): 478–486.
Malchaire, J., H. J. Gebhardt and A. Piette. 1999. Strategy for evaluation and prevention of risk due to work in thermal environments. *Annals of Occupational Hygiene* 43(5): 367–376.
Mattli, W. 2001. The politics and economics of international institutional standards setting: An introduction. *Journal of European Public Policy* 8(3): 328–344.
McNall, P. E., J. Jaax, F. H. Rohles, R. G. Nevins and W. Springer. 1967. Thermal comfort and thermally neutral conditions for three levels of activity. *ASHRAE Transactions* 73(1): 1–14.
Mishra, A. K. and M. Ramgopal. 2013. Field studies on human thermal comfort: An overview. *Buildings and Environment* 64: 94–106.
Nevins, R. G., F. H. Rohles, W. Springer and A. M. Feyerherm. 1966. A temperature-humidity chart for thermal comfort of seated persons. *ASHRAE Transactions* 72(1): 283–295.
Olesen, B. W. 1995. Ergonomics and international standards: History, organizational structure and method of development. *Applied Ergonomics* 26(4): 293–302.
Olesen, B. W. 2004. International standards for the indoor environment. *Indoor Air* 14(7): 18–26.
Olesen, B. W., F. R. d'Ambrosio Alfano, K. Parsons and B. I. Palella. 2016. The history of international standardization for the ergonomics of the thermal environment. In: *Proceedings of 9th Windsor Conference: Making Comfort Relevant*. Windsor, UK: Network for Comfort and Energy Use in Buildings, 15–38.

OSH Wiki. ISO standards in the area of ergonomics of the physical environment. https://oshwiki.eu/wiki/ISO_standards_in_the_area_of_the_Ergonomics_of_the_Physical_Environment (accessed on August 22, 2018).

Parsons, K. 1999. International standards for the assessment of the risk of thermal strain on clothed workers in hot environments. *Annals of Occupational Hygiene* 43(5): 297–308.

Parsons, K. 2001. Introduction to thermal comfort standards. http://www.utci.org/cost/publications/ISO%20Standards%20Ken%20Parsons.pdf (accessed on August 4, 2018).

Parsons, K. 2013. Occupational health impacts of climate change: Current and future ISO standards for the assessment of heat stress. *Industrial Health* 51: 86–100.

Parsons, K. C. 1995a. Ergonomics and international standards: Introduction, brief review of standards for anthropometry and control room design and useful information. *Applied Ergonomics* 26(4): 239–247.

Parsons, K. C. 1995b. International heat stress standards: A review. *Ergonomics* 38(1–2), 6–22.

Standardization: Fundamentals, Impact, and Business Strategy. 2010. APEC Sub Committee on Standards and Conformance Education Guideline 3 – Textbook for Higher Education. Singapore: Asia-Pacific Economic Cooperation (APEC).

Winslow, C. E. A., L. P. Herrington and A. P. Gagge. 1937. Physiological reaction of the human body to various atmospheric humidities. *American Journal of Physiology* 120: 288–299.

6 Who Creates Standards?

6.1 THE NATIONAL ORGANISATIONS FOR STANDARDISATION

Every country in the world has its own national organisation for standardisation. In Austria it is the ASI (acronym for Austrian Standards Institute (ASI) or germ. Österreichisches Normungsinstitut), in Belgium it is the NBN (acronym for Bureau voor Normalisatie/Bureau de Normalisation), in Finland it is the SFS (acronym for Suomen Standardisoimisliitto), etc.

Since there are 50 European national representatives involved in the work of the ISO (28 of them are officially members of the EU) and since there are also American organisations for standardisation, only a few will be presented here in the chapter.

The ISO and IEC account for about 85% of all international product standards (Mattli, 2001), but the IEC is not covered by this topic. Some Americans question the thesis that the ISO is the only truly international organisation, since the ISO is headquartered in Europe and since 1990 the majority of the working groups within the ISO technical committee are European based. Apart from perceiving the ISO standards as crafted from a European point of view, the USA producers also fear being left out of free trade using ISO international standards, since most American products are not certified as compliant with ISO standards, especially the ones falling in the scope of quality (Libicki, 1995). The disagreement between the European and American point of view on what makes an international standard truly international is whether the standard comes from an international organisation, in the sense that it has an international representation of national members and an international voting structure based on those national members, or by being internationally accepted and used by industry representatives (Mattli, 2001).

6.1.1 BSI Organisation

The BSI (acronym for British Standards Institution) was established in 1901 as the Engineering Standards Committee by Sir John Wolfe-Barry and was renamed the BSI in 1931. Today, it is a member body of the ISO together with other national bodies such as IRAM (Argentina), SARM (Armenia), SA (Australia), ASI (Austria), BELST (Belarus), NBN (Belgium), HZN (Croatia), DIN (Germany), AFNOR (France), etc. The BSI was Europe's first national standards body. It became part of the ISO after World War II.

The BSI committee met for the first time on January 22, 1901. One of the first standards to be published was related to the steel sections for tramways. During the 1920s, standardisation also spread to Canada, Australia, South Africa and New Zealand. World War II changed the course of the action, and ordinary work was stopped. Efforts were concentrated on producing over 400 war emergency standards. In 1979, the BSI proposed the UK's first management systems quality standard,

BS 5750, which later led to the ISO 9000 series of international standards. In 1996, the BSI published *BS8800* which inspired BS OHSAS 18001 *Occupational Health and Safety Management*, for achieving the suitable working conditions. In 1992, the world's first standard for environmental management systems BS 7750 *Specification for Environmental Management Systems* was published, which led to the creation of the ISO 140001 *Environmental Management Systems – Requirements With Guidance for Use* in 2015, the current international standard for environmental management systems (BSI website).

European Union Regulation 1025/2012 on European standardisation, which replaced parts of Directive 98/34/EC, places requirements on the BSI as a national standards body of an EU member state. The BSI makes available drafts and published national standards to other NSBs, European Standardisation Organisations and the European Commission on request and allows other NSBs to take part in the BSI's standards development work (EUR-Lex website).

6.1.2 DIN Organisation

DIN (acronym for the German Institute for Standardisation or germ. Deutsches Institut für Normung) is the independent organisation for standardisation in Germany. DIN was founded in 1917 as the Standards Association of German Industry (germ. Normenausschuss der Deutschen Industrie) and changed its name to its current form in 1975. From 1918, it has published the monthly journal *DIN Mitteilungen*, and in the same year a first German standard was published. In 1920, it was registered in Berlin as a non-profit organisation and became a part of the ISO in 1951. Experts from many fields bring their expertise, resulting in the creation of market-oriented standards and specifications that promote global trade, encouraging rationalisation, quality assurance and environmental protection as well as improving security and communication. DIN Standards are the result of work at a national, European and/or international level. Once the proposed standard draft is accepted, the project is carried out according to set rules of procedure by the relevant DIN Standards Committee, the relevant Technical Committee of the European standards organisation CEN or the relevant committee at the International Standards Organisation (ISO) (DIN website).

6.1.3 ASTM Organisation

The main mission of this American national organisation was the creation of uniform standards for the quality control of both raw materials and final products. Because of its negative experiences in attempting to establish a uniform system for the quality control and standardisation, C. B. Dudley suggested implementation of the technical committee system. These committees would be formed by representatives of the main parties, and the forums would be organised to display and discuss every aspect of specifications and testing procedures in quality control for a given material. The goal was to reach a consensus acceptable to both producers and to the customer, as well to the American railways. This negotiating principle became the basis for the formation of IATM (acronym for the International Association for Testing Materials) and nowadays the ASTM (acronym for American Society for Testing and Materials).

The technical committee of the American department of the IATM approved the first standard specifications for steel used in buildings constructed in 1901, under the title *Structural Steel for Bridges*. This standard was afterwards classified as ASTM's standard specification A7. Finally, in 1902, IATM was renamed ASTM. Parallel with the work of this organisation, the federal government of the USA tried to establish the NBS or the National Bureau of Standards in 1901. However, the founding of NBS was met with great resistance from both manufacturers and engineers, which disapproved of the government's plan to duplicate European practices and establish an authority body, which would force laws and regulations upon the industry. This decision proved to be crucial in the creation of the American democratic style of the development of standards and quality control. As the years passed, the global market climate in the USA changed. After the turn of the century, several new committees were formed quickened by the emersion of the new raw materials such as cement, clay, etc. In 1910, the ASTM introduced a yearbook, which later became the *Annual Book of ASTM Standards*. This publication became the foundational publication in the field of standard development. Each volume presented the complete work in the field of standardisation for the specific area, including the overview of all existing and revised standard specifications.

World War I introduced further changes and many companies reassigned their business to the military processing industry. In the early 1920s, the ASTM's main activities still focused on traditional materials such as the steel and cement. Over 100 technical committees were assigned to work on the supervision and development of standards in other areas. Over the following decade, American economic development began contributing to the rise of mass production in many industrial areas. One of the most propulsive areas was the automobile industry headed by H. Ford.

In 1981, the ASTM opened the first European standards distribution centre in London to broaden its activities in growing markets such as Japan, western Europe and so-called 'tiger economies' in the Pacific. Simultaneously, the fast development of new communication technologies was set, and international cooperation was improved. Today, the ASTM has broadened its participation in international waters forging close ties with national bodies from other countries. In 2001, the ASTM changed its name to ASTM International to reflect global participation in the creation of standards for worldwide usage (ASTM website). Currently, ASTM International operates in more than 140 countries and has published around 12,575 standards.

6.1.4 ASHRAE Organisation

The ASHRAE (acronym for the American Society of Heating, Refrigerating and Air Conditioning Engineers) was founded in 1959, and formed from the union of two organisations, the American Society of Heating and Air Conditioning Engineers (ASHAE) and the American Society of Refrigerating Engineers (ASRE). Today this organisation has over 50,000 members all over the world (ASHRAE website).

In the first decade of the twentieth century, these two associations had a somewhat different sphere of operation. In the 1920s, with the advancement of new technologies, air conditioning became a field of interest to both associations. Heating systems were slowly replacing steam and hot water heating with forced warm air systems, combing heating and air conditioning. In 1954, the American Society of Heating and Ventilating Engineers changed its name to the American Society of Heating and Air Conditioning Engineers. By the 1950s, a significant overlap was noted in the research programmes of both societies, so negotiations resulted in the merging of those associations into one unified organisation. The ASHRAE's prior mission was enhancement in the field of energy efficiency, air quality in interiors, industrial and building systems' sustainability.

An extremely important publication by the ASHRAE is the *Handbook of Fundamentals* (ANSI/ASHRAE Standard 55-2017). The origin of this publication goes way back to 1922 when the American Society of Heating and Ventilating Engineers published its *Heating and Ventilating Guide*. The guide was published annually until 1961, following their merging with the American Society of Refrigerating Engineers, and both societies started publication of the combined *ASHRAE Guide and Data Book*. Separate volumes were issued for *Fundamentals and Equipment* and *Applications*, and the *Guide and Data Book* was renamed the *Handbook of Fundamentals* in 1973, with separate *Systems, Applications* and *Equipment* volumes. Separate I-P and SI unit volumes were issued in 1985. Although volume groupings have shifted over the years, the compiled volumes have continued to become important literature in the field of heating, refrigerating and air conditioning (ASHRAE website).

6.2 THE INTERNATIONAL ORGANIZATION FOR STANDARDIZATION (ISO)

The most important international organisation for standardisation is the ISO. The twentieth century has been marked by accelerated evolution and the application of new materials and their testing methods and technological advancement. Simultaneously, the idea of the need for a clear quality management system implementation for raw materials and testing procedures has been developed. This resulted in the development of the first detailed publications and reports on materials and their specifications. The clear quality management system is a primary requirement which aims for consumer protection and gives precise guidance to producers while designing new products and testing procedures.

The acronym ISO is derived from the Greek *isos*, meaning equal and an arrangement was made to accept this acronym as the official abbreviation of the organisation (ISO website – about us). Since establishment in 1947 with an initial 67 technical committees, today's members are from 162 countries, 783 technical bodies and 3368 institutional bodies; this international organisation for standardisation successfully operates, with headquarters in Geneva, in the field of publishing standards from many different areas (agriculture, building construction, engineering, medical equipment, etc.). Only one national referent body presents every country. Over its history, the ISO has developed more than 14,000 standards. Though this international institution initially focused on developing technical standards for specific products such as films and screws, the ISO has expanded its scope significantly, and now develops management system standards and other protocols that have significant environmental and social policy implications (Morikawa and Morrison, 2004).

The ISO was originally formed from the union of two organisations, the International Federation of National Standardizing Associations (ISA), established in 1926, and the United Nations Standards Coordinating Committee (UNSCC), established in 1944. Although the ISA was established in New York, the standardising body of the United States never participated in its work (Olesen et al., 2016).

Four years after the ISA was dissolved in 1942, delegates from 25 countries decided to create a new international organisation, the ISO, to facilitate international coordination and unification of industrial standards (Morikawa and Morrison, 2004).

In 1947, after World War II was over, the ISO came into existence. By the end of 1947, the ISO had been granted Consultative Status by the United Nations and work began to establish links with the many international organisations (Olesen et al., 2016).

The first ISO standard (called *Recommendations*) was published in 1951 under the name ISO/R 1:1951 *Standard Reference Temperature for Industrial Length Measurements* (ISO/R 1:1951). In 1975, it was published as ISO 1:1975 *Standard Reference Temperature for Industrial Length Measurements* (ISO 1:1975) and in 2002 renamed ISO 1:2002 *Geometrical Product Specifications (GPS) – Standard Reference Temperature for Geometrical Product Specification* (ISO 1:2002). The 2002 version has been revised and replaced by the 2016 version (ISO 1:2016). The current version defines the concept of reference temperature and standard reference temperature. It also specifies the standard reference temperature value for specification of geometrical and dimensional properties of an object such as size, location, orientation, form and surface texture. Nowadays, the ISO publishes about 1000 international standards per year.

The official *ISO Journal* started with monthly publication in 1952 and in 1960 the ISO published a standard in correspondence to the international system of units (SI) called ISO 31 *On Quantities and Units* (ISO 31:1960), which has later been replaced by ISO 80 000 (ISO 80000-1:2009).

In 1987, the ISO published the first quality management standard called ISO 9002: 1987 *Quality Systems – Model for Quality Assurance in Production and Installation* (ISO 9002:1987). This standard was developed on the basis of the British quality standard (BS 5750-2:1987, EN 29002:1987). Subsequently, a whole ISO 9000 series was published, such as ISO 9000:2015 *Quality Management Systems – Fundamentals and Vocabulary* (ISO 9000:2015), ISO 9001:2015 *Quality Management Systems – Requirements* (ISO 9001:2015) and ISO 9004:2009 *Managing for the Sustained Success of an Organisation – A Quality Management Approach* (ISO 9004:2009) superseded by ISO 9004:2018 *Quality Management – Quality of an Organisation – Guidance to Achieve Sustained Success* (ISO 9004:2018), etc. It is also important to mention that from 1979 the work of the ISO is completely coordinated with requests set by *The Technical Barriers to Trade* (TBT) *Agreement* with the *World Trade Organisation* (WTO).

Today, the ISO operates as an independent, non-governmental organisation. There are three types of member – full members, correspondent members and subscriber members. The full members influence the standards and strategy development by participating and voting in ISO technical and policy meetings. They also sell and adopt ISO international standards nationally. Correspondent members only observe the development of the standards and strategies by attending ISO technical and policy meetings as observers while the subscriber members cannot participate in the ISO's work nor sell or adopt ISO international standards nationally.

The ISO has a specific structure. On top, there is the General Assembly, which is the overarching organ and ultimate authority of the ISO. Underneath, there is the Council, which is the core governance body of the ISO and reports to the General Assembly. The Council has direct responsibility for a number of bodies such as the President's committee, the Council Standing committees, the advisory groups, the Committee on Conformity Assessment (CASCO), the Committee on Consumer Policy (COPOLCO) and the Committee to Support Developing Countries (DEVCO). The management of the technical work is taken care of by the Technical Management Board (TMB), which also reports to the Council. This body is also responsible for the technical committees that lead standard development and any strategic advisory boards created on technical matters (ISO website-structure), Figure 6.1.

Members of ISO committees (participating and observer members) are delegates of national standards bodies. The members of a country's delegation are drawn from its national mirror committee, and they represent the views of their national members on the international stage. In contrast, members of working groups are experts who have been nominated by their national standards body and should be aware of their national point of view but act in a personal capacity. Experts are appointed by the participating members of the parent technical committee (*Getting Started Toolkit for ISO Working Group Convenors*, 2017).

There is more than 180 technical committees (TCs), around 700 subcommittees (SCs) and about 1500 working groups (WGs) within the ISO, which carried out all of the technical work (Jüptner, 1984).

The ISO's standards development process consists of three main phases. In the first phase, the need for a standard is expressed, typically by an industry sector, which communicates this need to a national member body within the ISO. The member

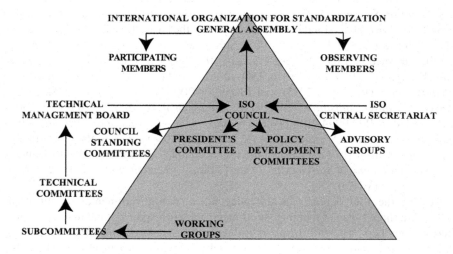

FIGURE 6.1 The hierarchy of ISO.

body then carries the new work item proposal to the ISO. Once ISO member bodies formally agree by ballot to the need for an international standard, technical experts from countries interested in the subject matter finalise the technical scope of the future standard. Countries then proceed to negotiate the detailed contents and provisions within the standard, using a consensus-based approach. The third phase comprises the formal approval of the final draft document, at which time the agreed-upon text is published as an ISO international standard (Morikawa and Morrion, 2004).

The lead standard development is performed by the ISO technical committees. As soon as a draft has been developed and accepted by the technical committee, it is shared with the ISO's member bodies who vote on it. A publication which will become an international standard should be approved by 75% of the voting membership bodies.

The development of a particular standard is the responsibility of a single technical committee while the final working draft is assigned to either a subcommittee or a working group of the parent committee. When there is a significant amount of relatively unrelated work being undertaken, the work is divided through separate, semi-autonomous subcommittees. The subcommittees make their own decisions, such as approving new work item proposals, approving committee drafts, establishing and disbanding working groups and so on. Each technical committee and subcommittee has its own chairman and secretary. Underneath each subcommittee, there is a number of working groups in which related work items are developed. The drafting of standards takes place in project groups, which are usually grouped into working groups that have responsibility for a particular aspect of the work of the technical committee or subcommittee (Hatto, 2001).

A standard that has been developed at an international level can be simultaneously adopted as a European standard by means of parallel voting procedures and automatically adopted by the national standards organisations (ISO website: about us).

The international standards, which describe the terms, the measuring methods, the thermal environments and the other parameters connected to the transfer and effects of heat, belong to the category of ISO standards in the field of the ergonomics of thermal environments (Špelić, 2016). Supporting standards on textile and clothing measuring methods fall into the scope of the ISO/TC 38 (ISO website-48148). This will be covered in detail in Chapters 7 and 9.

REFERENCES

ANSI/ASHRAE Standard 55-2017 Thermal Environmental Conditions for Human Occupancy. 2017. Atlanta, USA: American Society of Heating, Refrigerating and Air Conditioning Engineers Inc. (ASHRAE), ISSN: 1041–2336.

ASHRAE. http://www.ashrae.org (accessed on October 10, 2012).

ASTM. http://www.astm.org (accessed on April 13, 2013).

BS 5750-2:1987, EN 29002-1987 Quality Systems. Specification for Production and Installation. 1987. BSI.

BSI. http://www.bsigroup.com/ (accessed on December 5, 2017).

DIN. DIN and our partners. https://www.din.de/en/din-and-our-partners (accessed on May 13, 2015).

European Parliament. 2012. *Official Journal of the European Union.* http://eur-lex.europa.eu/LexUriServ/LexUriServ.do?uri=OJ:L:2012:316:0012:0033:EN:PDF (accessed on July 15, 2015).

Getting Started Toolkit for ISO Working Group Convenors. 2017. ISO-International Organization for Standardization.
Hatto, P. 2001. *Standards and Standardisation, A Practical Guide for Researchers.* European Commission, Directorate-General for Research & Innovation, Directorate G – Industrial Technologies. https://ec.europa.eu/research/industrial_technologies/pdf/practical-standardisation-guide-for-researchers_en.pdf (accessed on August 18, 2018).
ISO. About us. http://www.iso.org/about-us.html (accessed on October 12, 2012).
ISO. ISO/TC 38 textiles. https://www.iso.org/committee/48148.html (accessed on August 23, 2018).
ISO. Structure and governance. https://www.iso.org/structure.html (accessed on October 12, 2012).
ISO 1:1975 Standard Reference Temperature for Industrial Length Measurements. 1975. ISO-International Organization for Standardization.
ISO 1:2002 Geometrical Product Specifications (GPS) – Standard Reference Temperature for Geometrical Product Specification and Verification. 2002. ISO-International Organization for Standardization.
ISO 1:2016 Geometrical Product Specifications (GPS) – Standard Reference Temperature for the Specification of Geometrical and Dimensional Properties. 2016. ISO-International Organization for Standardization.
ISO 31:1960 Quantities and Units. 1960. ISO-International Organization for Standardization.
ISO 80000-1:2009 Quantities and Units – Part 1: General. 2009. ISO-International Organization for Standardization.
ISO 9000:2015 Quality Management Systems – Fundamentals and Vocabulary. 2015. ISO-International Organization for Standardization.
ISO 9001:2015 Quality Management Systems – Requirements. 2015. ISO-International Organization for Standardization.
ISO 9002:1987 Quality Systems – Model for Quality Assurance in Production and Installation. 1987. ISO-International Organization for Standardization.
ISO 9004:2009 Managing for the Sustained Success of an Organization – A Quality Management Approach. 2009. ISO-International Organization for Standardization.
ISO 9004:2018 Quality Management – Quality of an Organization – Guidance to Achieve Sustained Success. 2018. ISO-International Organization for Standardization.
ISO/R 1:1951 Standard Reference Temperature for Industrial Length Measurements. 1951. ISO-International Organization for Standardization.
Jüptner, H. 1984. ISO standards in the field of ergonomics: A report of ISO/TC 159 activities. *Applied Ergonomics* 15(3): 211–213.
Libicki, M. C. 1995. What standards do? Chapter 1. In: *Information Technology Standards: Quest for the Common Byte.* Burlington, MA: Butterworth-Heinemann, Elsevier Inc., ISBN: 1-55558-131-5.
Mattli, W. 2001. The politics and economics of international institutional standards setting: An introduction. *Journal of European Public Policy* 8(3): 328–344.
Morikawa, M. and J. Morrison. 2004. *Who Develops ISO Standards? A Survey of Participation in ISO's International Standards Development Processes.* Oakland, CA: Pacific Institute for Studies in Development, Environment, and Security. https://pacinst.org/reports/iso_participation/iso_participation_study.pdf (accessed on August 18, 2018).
Olesen, B. W., F. R. d'Ambrosio Alfano, K. Parsons and B. I. Palella. 2016. The history of international standardization for the ergonomics of the thermal environment. In: *The Book of Proceedings of the 9th Windsor Conference: Making Comfort Relevant*, eds. L. Brotas, S. Roaf, F. Nicoland and M. Humphreys. Windsor, UK: Network for Comfort and Energy Use in Buildings (NCEUB), 15–38, ISBN: 978-0-9928957-3-0.

Špelić, I., D. Rogale and A. Mihelić-Bogdanić. 2016. An overview of measuring methods and international standards in the field of thermal environment, thermal characteristics of the clothing ensembles and the human subjects' assessment of the thermal comfort. *Tekstil* 65(3–4): 110–122.

7 The Standardisation of Thermal Comfort

7.1 THE HISTORY OF STANDARDISATION

The era of the Industrial Revolution brought the development of new materials, the need for implementation of clear specifications and the establishment of standards. This was a time of emerging of new materials and their application in technological advancement since a vast number of different materials were being developed and applied, especially the steel for railroads. Simultaneously, with the introduction of new materials and invention of new techniques, need for the producers to prevent the eventual application of the unsuitable raw materials have indicated explicit quality system management. During this period, the first detailed documents with material and resources specifications appeared on American soil, along with equipment for quality determination, despite the resistance demonstrated by the American public.

The absence of standardised specifications and testing procedures was fertile soil for diversity in determining design requirements, performance criteria, material qualification and certification among different producers. This left potential buyers quite unsure on the uniformity and quality of the steel rails for railroads and frequently gave them a reason to complain (ASTM website, history). Since standards are a set of specifications or guidelines to ensure that a product, service or process is compliant with the highest quality guidelines, they are used as tools to enhance safety, improve productivity, protect the environment and create technology. Standards help by identifying the optimum parameters for the performance of a process, product or service and the method for evaluating product conformity and quality. The idea behind the first specification was to ensure that both the needs of the producers and the users were met. It also initiated the development of unified standard methods of testing, and the research of important material properties and apparatus used for this purpose.

The first standards bodies were established in the early years of the twentieth century, both on a national and on an international level. In 1901, several British engineering societies founded the Engineering Standards Committee (later the British Engineering Standards Association). In 1906, the International Electrotechnical Commission (IEC) was established (IEC website, history). The IEC developed many of the techniques and institutional mechanisms that laid the procedural foundations for other bodies. The American Engineering Standards Committee (AESC) was established as the first American general standards-setting body in 1918. The International Federation of National Standardizing Associations (ISA) followed in 1926 to create standards in the field of mechanical engineering and should be considered as the first international standards body. By 1941, both the IEC and the ISA had temporarily stopped their work due to a break in the world's economy after

World War I and the beginning of World War II. The ISA was superseded by a new organisation, the International Organization for Standardization. Together with the IEC, which resumed its operation after World War II, and the International Telecommunication Union (ITU – grown from the International Telegraph Union), the ISO is one of the three principal international standards bodies today (Morikawa and Morrison, 2004; Wenzlhuemer, 2010).

7.2 THE BEGINNINGS AND FIRST STANDARDS OF THERMAL COMFORT

In the United States, C. B. Dudley initiated material testing for American railroad construction and issuing procedures for standardising specifications. He published his first major report in 1878 under title *Chemical Composition and Physical Properties of Steel Rails*, in which he analysed the durability of different types of steel. With his work, he continued to implement the idea of collaboration between steel producers and manufacturers, which resulted in the creation of the unitary quality control system and also establishment of the ASTM organisation (American Society for Testing and Materials) in 1898. In 1901, the BSI (British Standards Institution) was also formed. This places the ASTM and BSI as one of the first national organisations for standardisation (ASTM website, history, Špelić et al., 2016).

There was a general shift in the 1960s away from producing international standards covering basic test methods and terminologies towards producing standards related to the performance, safety and health aspects of particular products (*Friendship Among Equals*, 1997). Thermal comfort and thermal environments generally were brought into the spotlight during the twentieth century as working environments became more and more important.

But designing working conditions wasn't the only important matter. Warfare, such as World War I (WWI), World War II (WWII), the Korean War and the Cold War, with soldiers being exposed to harsh climates, pressed scientists to conduct studies for preventing thermal stress (Olesen et al., 2016). Research in the field of the ergonomics of the thermal environment progressed quickly due to findings on human physiology and unfortunately due to warfare. During WWI, personal equipment was inadequate and too heavy. It was a common practice for soldiers to discard clothing and equipment because the standard individual load was about two-thirds of a typical soldier's body weight. Shoes were not waterproof, and 90% of the soldiers suffered from cold injuries during the harsh winter of 1917/1918. These problems were not solved even when the U.S. Army entered WWII (Friedl and Santee, 2012). The inadequately designed cold weather uniforms became a source of great dissatisfaction and loss. This prompted researchers to design both new uniforms, personal equipment and develop new measuring protocols for acquiring a satisfactory thermal state for the soldiers. During WWII, the soldiers were also exposed to a wider range of climatic conditions, from desert conditions, moist heat, to cold.

New transportation means, such as airplanes, moreover exposed humans to hazards, especially the environmental ones. In the early days of aviation, flying airplanes was fraught with many hazards and difficulties. The open aircraft of that era usually caused pilots and passengers great discomfort with their deafening roar, splattering oil and icy wind, not to mention the lashing rain or snow (Sweeting, 2015).

On American soil, the first studies on physiology were in the 1920s. Those early studies on physiology were motivated by working performance. From the late 1920s to 1940s, Harvard Medical School studied the effects of fatigue on workers in the Harvard Fatigue Laboratory. Afterwards, the focus shifted to applied physiology since the U.S. Army entered WWII. The Harvard Medical School, the Massachusetts General Hospital, the U.S. Army Medical Armored Research Laboratory and the U.S. Army Quartermaster Climatic Research Laboratory collaborated to support U.S. Army troops in the field. Most of the studies included the testing of the sleeping bags. The U.S. Army Medical Armored Research Laboratory tested the problems of armoured vehicles during desert operations, while the U.S. Army Quartermaster Climatic Research Laboratory performed research on the problems of cold weather operations in Northern Europe and Alaska (Goldman, 2006).

At the beginning of the twentieth century, Adolph Pharo Gagge (1908–1993) was a pioneer in researching the interactions of the human body with the environment.

In 1930s, Gagge joined the independent research institute the John B. Pierce Laboratory of Physiology and Health in the Modern Environment, New Haven, Connecticut, affiliated with Yale University and formally known as the John B. Pierce Laboratory of Hygiene. The John B. Pierce Laboratory is a non-profit, independent research institute was founded in 1933. In the 1930s the work of C. E. A. Winslow, L. P. Herrington and A. P. Gagge defined the physical principles that are central to how organisms exchange energy with their environment. Thermal exchange was measured in terms of four basic physical processes: radiation, convection, vaporisation and conduction. These pioneering studies on human calorimetry laid the foundation for later studies in pulmonary and thermal physiology and air quality and pollution (the John B. Pierce Laboratory website). After publishing (Gagge, 1936) his thermal balance model in 1936, the so-called 'two-node' model was born, and the studies on the human thermal comfort system started.

During WWII and the Korean War, the U.S. Army also suffered severe human losses due to extreme weather conditions. Over 90,000 U.S. Army and Army Air Force casualties during WWII were attributable to cold injury. The German Army cold-injury casualties were at least as high. Another 10,000 casualties resulting from cold injury occurred during the Korean War. Currently, cold-injury prevention is an area of major command emphasis for army units operating in cold climates (Young et al., 1996).

During WWII, frontline infantrymen incurred 87% of all general cold-induced injuries, and in many cases the combat effectiveness of entire infantry units was nullified (Endrusick et al., 1992). Most of the research in that period concentrated around military performance. In the 1940s and 1950s, research in the B. Pierce Laboratory centred entirely on the military efforts for WWII and the Korean War (the John B. Pierce Laboratory website). During that time, the John B. Pierce Laboratory of Hygiene and Natick's Base laboratories carried a series of studies on the human body's response to extreme conditions and protection against thermal stress. Studies that started during that time led to the definition of the most important indices for cold and hot environments (Goldman, 2006). The testing on human subjects, for either hypothermia protection or heat-related injuries, is not without risk. The in vivo testing on humans also provides fewer options, since life is put at risk. So mathematical modelling and apparatus development, such as thermal manikins, was a simpler but more practical solution.

The first thermal testing began in the United States. By November of 1942, the science had advanced considerably. The human subject measurements now included a 9-point mean skin temperature, measured every 15 minutes, the rectal temperature measured hourly and body heat production (respiratory O_2 consumption) measured continuously. During that time, the first model of an 'electrically heated dummy' with an internal fan blowing air across internal heating coils throughout the torso and down the extremities, using an on/off thermostatic control to simulate human skin temperature, was introduced. This is considered to be the forerunner of modern heated manikins, and it was constructed to measure the performance of sleeping bags for use by military personnel. But it had some shortcomings so by April of 1943 an improved 'electrical dummy', with an electrical fan blowing air through tubes to the extremities, was built (Goldman, 2006).

Up to September of 1943, the Harvard Fatigue Laboratory had improved physiologic measurements. The 11-point mean weighted skin temperature was introduced, using weightings proportional to the surface area involved, rectal temperature measurement was standardised at a depth of 15 cm and plotted every 15 minutes, and the calculation of mean body temperature was corrected. A wide variety of clothing materials were being measured on a new Cleveland heated flat plate, guarded on all four sides, and across the bottom, by heated sections set at identical temperatures as the central test section. This is still the recommended technique for measuring material insulation, but cannot simulate the effects of material weight or garment drape, size, cut, closures, etc. A new heated hand and foot model was constructed to measure thermal insulation for hand- and footwear. This new electrically heated dummy, made of 1.6 cm thick sheet copper, was heated internally by resistance wire and electric light bulbs, with an internal fan and on/off thermostatic control of temperature. A difference in measured insulation values of 5% was now considered significant. Based on the success of using the manikins to supplement physiological determinations, e.g. the loss of clothing insulation worn by marching soldiers in the cold calculated from heat-balance equations (Goldman, 2006).

In 1946, the first electrically heated manikin made of copper, called Chauncy, was produced from copper at Harvard Fatigue Laboratory (Boston) and U.S. Army Research Institute for Environmental Medicine (USARIEM, Natick, Massachusetts), mainly through the work of Dr Harwood Belding. The head, thumbs and forward part of the feet were made removable, and the shoulders were articulated to rotate 180 degrees, both for internal access and to facilitate donning clothing, hand and footwear. Thermocouple temperature sensors, placed in 22-calibre bullet cartridge cases inserted into the skin at appropriate sites, were used to measure the average skin temperature of each section of the manikin, whose skin was blackened to more closely approximate the thermal emissivity of human skin. It had connectors for power, thermostat control and skin temperature sensors in the area of what would have been the human navel. Afterwards, more manikins had been produced and shipped to the U.S. Navy in Philadelphia, to the American Society of Heating and Ventilating Engineers (ASHVE) Laboratory, which later moved to Elizabeth McCullough at Kansas State University, to A. P. Gagge at the J. B. Pierce Foundation Institute, three of them were shipped to the U.S. Army Air Force at Wright Patterson Air Force Base, and the original Chauncy was taken by Dr Harwood Belding to the University of Pittsburgh, Pennsylvania School of Public Health. The manikins were later produced in a variety of forms from the water immersion manikin, standard electrically heated manikin and the sweating manikin (Goldman, 2006).

In the 1950s and 1960s, scientists from the J. B. Pierce Laboratory pioneered studies of building regulations and the physiological effects of the thermal environment. Professional organisations such as the American Society for Heating, Refrigeration and Air Conditioning Engineering (ASHRAE) interacted closely with the J. B. Pierce Laboratory in setting industry standards for human comfort and health. Under the directorship of James D. Hardy, a renowned scientist in the fields of pain, thermal sensation and body temperature regulation, the J. B. Pierce Laboratory came to place greater emphasis on the basic biological processes that underlie responses to environmental stimuli, which is a hallmark of current research at the Laboratory (the John B. Pierce Laboratory website).

During 1955, the U.S. Army Quartermaster Climatic Research Laboratory was renamed the Environmental Research Division and moved to Natick. In the 1960s, the U.S. Army Research Institute of Environmental Medicine was created by merging the parts of the Environmental Research Division with the U.S. Army Medical Armored Research Laboratory (Goldman, 2006).

In the course of the Cold War, during the early 1960s, the question of the ability of U.S. troops to operate with old cold-protective ensembles developed during WWII was raised, and the first sweating manikin was produced (Goldman, 2006).

From the 1960s onwards, many researchers like Jan A. J. Stolwijk and James D. Hardy from the John B. Pierce Laboratory at Yale University (Stolwijk and Hardy, 1966a, 1966b; Stolwijk, 1969, 1970, 1971), Frederic H. Rohles Jr. and Ralph G. Nevins from the Institute of Environmental Research at Kansas State University (Rohles and Nevins, 1971; Rohles et al., 1973), Rohles and M. A. Johnson (Rohles and Johnson, 1972), Baruch Givoni, Ralph F. Goldman and K. B. Pandolf from the U.S. Army Institute of Environmental Medicine (Givoni and Goldman, 1972, 1973; Pandolf et al., 1977), Adolph P. Gagge and Yasunobu Nishi from the John B. Pierce Laboratory at Yale University (Gagge, 1936, 1937; Nishi and Gagge, 1970a, 1970b; Gagge et al., 1971, 1986; Nishi and Gagge, 1971, 1973, 1974, 1976, 1977; Gonzales et al., 1974), Paul A. Bell (Bell, 1980, 1981; Bell and Greene, 1982), Michael A. Humphreys from the Oxford Centre for Sustainable Development and J. Fergus Nicol from the National Institute of Medical Research (Humphreys and Nicol, 1998), Richard J. de Dear, Gail S. Brager and Charlie Huizenga et al. from the Center for the Built Environment at the University of California (de Dear and Brager, 1998a, 1998b; Huizenga et al., 2001) performed studies on thermal comfort and the human thermal regulation deriving the two- and the multi-node models. Jan A. J. Stolwijk from the John B. Pierce Laboratory developed the multi-node model (Stolwijk, 1970, 1971) by dividing the body into six segments and each segment into four layers in a radial direction. On the basis of the Stolwijk–Hardy multi-segmented model, the Fiala and UC Berkeley thermal comfort multi-segmented multi-node models were developed (Tanabe et al., 1995; Fiala et al., 1999, 2012; Huizenga et al., 2001; Bröde et al., 2013). Most of the studies described above were the basis for creating the official unified international standards.

In America, the ASHRAE organisation was involved in research on comfort studies as early as the 1950s at the Institutes for Environmental Research, although the origins can be traced to much earlier when in 1894 the ASHVE was established, following the establishment of the ASHVE Bureau of Research in 1919. The ASHVE, later to become a part of ASHRAE, continued his cooperation with John Bartlet Pierce. John B. Pierce provided funds for establishing the John B. Pierce Foundation for technical research in heating, ventilating and sanitation, and later founded the laboratory under his name (Janssen, 1999). ANSI/ASHRAE Standard 55 *Thermal Environmental Conditions for Human Occupancy* is the first standard in the field of the ergonomics of the thermal environment and was released in 1966 (ANSI/ASHRAE Standard 55-1966; Brager and de Dear, 2000). ASHVE research led to a comfort chart that correlated temperature, humidity and comfort responses. It was first published in the *ASHVE Guide* in 1924, and it continued to be published in the guide until it was replaced by ANIS/ASHRAE Standard 55-1974 *Thermal Comfort*. The comfort chart was modified to reflect the responses of clothing, heating/cooling system designs, and living habits in the 1981 and 1992 edition of the same standard (Janssen, 1999).

ASHRAE Standard 55 was the first standard to establish acceptable thermal comfort ranges for indoor spaces and the first to summarise the different parameters that affect human comfort (temperature, relative humidity, airspeed and occupant activity and clothing insulation). From its first publication onward, the ANSI/ASHRAE 55 standard went through many changes, incorporating new findings. The standard was republished in 1974, 1981, 1992, 2004, 2004 and the latest edition was released in 2017 (ANSI/ASHRAE Standard 55-2017). The 1992 version allowed modest increases in operative temperature beyond the PMV-PPD model limits as a function of airspeed and turbulence (ANSI/ASHRAE Standard 55-1992). Fanger's Predicted Mean Vote (PMV) and Predicted Percent Dissatisfied (PPD) model was the first adaptive thermal comfort model which was widely accepted (de Dear et al., 2013).

During the 1970s and 1980s, Povl Ole Fanger (1934–2006) defined the indexes of comfort and revised Gagge's two-node model (Fanger, 1970, 1973a, 1973b, 1982; Fabbri, 2015). Fanger is also known for determining the six basic parameters which affect human thermal comfort (Fanger, 1970). The environmental variables that affect human responses to thermal environments are air temperature, radiant temperature, humidity and air movement, while the metabolic heat generated by human activity and the clothing worn by a person is identified as personal factors.

After Fanger published his thermal comfort model (the PMV-PPD model) (Fanger, 1970), both *ASHRAE 55* and *ISO 7730* adopted his research. However, in 1970 (Fanger, 1970), Fanger presented the comfort equation originally under the British imperial system of units. The SI version was afterwards presented in the work of Olesen, *ASHRAE 55*, *ASHRAE Handbook of Fundamentals* and *ISO 7730* (Olesen, 1982; *ASHRAE Handbook of Fundamentals*, 2017; ANSI/ASHRAE Standard 55-2017; ISO 7730:2005).

Fanger's PMV model is based on human body heat-balance considerations. Taking into account that thermal sensation is influenced by environmental (air temperature, radiant temperature, humidity and air velocity) and personal factors (activity and clothing), the PMV index predicts thermal comfort on a 7-point sensation scale. The Predicted Percentage of Dissatisfied people index, derived from the PMV index, predicts the percentage of a large group of people likely to be thermally (dis-)satisfied with their environment. The evaluation of the PMV index is not easy, since many of the parameters have to be estimated or require sensing modalities that may not be available. For this reason, both ASHRAE 55 and ISO 7730 introduce simplified calculation methodologies to define acceptable limits for thermal comfort (Kontes et al., 2017).

The problem of defining thermal comfort conditions in naturally ventilated environments has led to the creation of the thermal comfort Adaptive Model since Fanger's model was only applicable in artificially controlled spaces. The thermal comfort Adaptive Model was first introduced by Humphreys and Nicol in the 1970s (Nicol and Humphreys, 1972) and due to de Dear and Brager it was incorporated into ASHRAE Standard 55 in 2004 (de Dear et al., 1997; de Dear and Brager, 1998a, 1998b, 2002). They proposed revisions to an older version of ASHRAE Standard 55 (ANSI/ASHRAE Standard 55-1992). One of the more contentious theoretical issues in the applied research area of thermal comfort has been the dialectic between 'adaptive' and 'static' models (de Dear et al., 1997; Bell, 2004).

R. J. de Dear and G. S. Brager claimed the standards ASHRAE 55:1992 (ANSI/ASHRAE Standard 55-1992) and older version of ISO 7730:1994 (ISO 7730:1994) to establish relatively tight limits on recommended indoor thermal environments without distinguishing the difference between what would be considered thermally acceptable in buildings conditioned by natural ventilation versus air conditioning (Brager and de Dear, 2000).

Their research (ASHRAE RP-884) demonstrated that occupants of naturally ventilated buildings are comfortable in a wider range of temperatures than occupants of buildings with centrally controlled HVAC systems (Brager et al., 2004).

The 2004 edition introduced significant changes like the adaptation of the Computer Model Method for general indoor application that made agreements to ISO 7726 and ISO 7730. It also introduced the Adaptive Method which relied on research to support natural ventilation designs for more sustainable, energy efficient and occupant-friendly designs (ANSI/ASHRAE Standard 55-2004).

According to ASHRAE 55-2004, indoor conditions are to be in accordance with the PMV-PPD model, except where the Adaptive Method can be applied with its less restrictive and sliding range of thermal limits. The refinement in ASHRAE 55 is to clarify the evaluation required to test against the Adaptive Method's limits on outdoor temperatures, as well as the prohibition on its use when mechanical cooling systems are provided for the space (Turner, 2008).

In 2010, more changes were accepted into the new version of the ANSI/ASHRAE 55 standard. The 2010 edition of the standard clarified that the upper humidity limit shown on the psychrometric chart in the Graphic Comfort Zone Method applies only to that method. Higher humidity limits are allowed if evaluated with the Computer Model Method and no limits are imposed on the Adaptive Model. It also revised requirements and calculation methods when increased air movement is used to maintain comfort in warm conditions. Standard Effective Temperature (SET) is reintroduced into the standard as the calculation basis for determining the cooling effect of air movement, and the calculation method has been simplified (ANSI/ASHRAE Standard 55-2010).

2017 version of the ANSI/ASHRAE 55 standard clarified the three comfort calculation approaches (a graphic method for simple situations; an analytical method for more general cases; and a method that uses elevated air speed to provide comfort). It also includes a simplification of Appendix A to a single procedure for calculating operative temperature, clearly stated requirements and calculation procedures that appear sequentially, an update to the scope to ensure the standard is not used to override health and safety, and critical process requirements (ANSI/ASHRAE Standard 55-2017).

Although standards like ASHRAE 55-2017 are a major carrier for spreading the adoption of adaptive comfort models among practitioners and designers, they still are affected by a few uncertainties in the application that prevent their full exploitation. The issue of their application in hybrid or mixed-mode buildings has to be solved, and a new approach should be identified to set a rule for univocally identifying the period when the adaptive comfort models can be applied, because of their intermittence compliance to the current rules during the transition seasons and in the hottest days of the year (Bai et al., 2017).

On an international level, the standards defining thermal environments and methods to deduce certain strain indices were adopted almost 15 years later (Olesen et al., 2016). The first international standards in ergonomics were formally considered in 1973, when the symposium by the International Ergonomics Association (IEA) was held at Loughborough University of Technology (UK), to discuss the international standards in the field of ergonomics. The recommendations led to the establishment of the ISO technical committee ISO/TC 159 *Ergonomics*, with Deutsches Institut für Normung (DIN) holding the secretariat and six subcommittees (SC). These have been condensed into the present four subcommittees: SC1 *Ergonomics Guiding Principles*, later renamed *General Ergonomics Principles*, SC3 *Anthropometry and Biomechanics*, SC4 *Ergonomics of Human-System Interaction* and SC5 *Ergonomics of the Physical Environment*. There are over 15 working groups (WG) working to produce international standards under those four corresponding subcommittees, established during 1974 (Jüptner, 1984; Parsons, 1995a; 2001; Parsons et al., 1995). Today under the domain of the ISO technical committee ISO/TC 159 *Ergonomics*, there are around 132 standards published and around 24 under development, with 20 participating members and 14 observing members (ISO website). In 1974, subcommittee ISO/TC 159 SC5 *Ergonomics of the Physical Environment* was established, but it wasn't until 1979, when the subcommittee for the ergonomics of the physical environment SC5 formed working group WG1 accounting for the ergonomics of the thermal environment, that the international work on standards in this area started. The working group ISO/TC 159/SC5/WG1 ergonomics of the thermal environment is the prime organisational domain to work on standards in the area of the thermal comfort and thermal environments (Olesen et al., 2016).

The subcommittee ISO/TC 159/SC5 *Ergonomics of the Physical Environment* develops international standards in the area of the ergonomics of the physical environment. There are 33 standards published under this subcommittee and three currently under revision (ISO 7933 currently as a draft international standard, ISO 8996 currently as committee draft and ISO 10551 currently as a draft international standard) (ISO website). This subcommittee consists of three working groups, one of which is WG1 which covers the thermal environments. There are around 20 international standards currently in usage that has been created by this working group (Parsons, 2000a; 2018).

In the 1980s, European and international standards began developing (Parsons, 1995b). Since the ISO formed the working group ISO/TC 159/SC5/WG1 for the area of the ergonomics of the thermal environment, a series of the international standards has been released. ISO 7243 *Hot Environments – Estimation of the Heat Stress on the Working Man, Based on the WBGT Index (Wet Bulb Globe Temperature)* was the first European standard in the field of the ergonomics of the thermal environment and was released in 1982 (ISO 7243:2017; BS EN 27243:2017). ISO work is divided among more than 250 technical committees, and each technical committee consists of several working groups, which will be described below when discussing the ISO.

Regarding clothing research for usage in different environmental conditions, commercial application was even slower, for either every day or protective clothing.

The first standards based on thermal manikin measurements were developed in 1986 when four standards for thermal insulation measurements of cold-protective clothing were proposed and accepted by Denmark (Holmér, 2006). The first one was INSTA 352 and covered definitions, units and symbols, (INSTA 352, 1986); the second INSTA 353, covered the testing procedure of the protective clothing with thermal manikins (INSTA 353, 1986). The third and fourth proposals specified conditions and procedures for the measurement of a single piece of a garment (INSTA 354, 1986) and a complete clothing ensemble (INSTA 355) (Holmér, 2006).

The two EU Directives on personal protective equipment published in 1989, dealt with minimal safety and health requirements for the use of personal protective equipment.

One of them was Directive 89/656/EEC – *Use of Personal Protective Equipment* adopted on November 30, 1989 by the European Council (EU-OSHA website). This Directive lays down minimum requirements for personal protective equipment (PPE) used by workers at work. Personal protective equipment must be used when the risks cannot be avoided or sufficiently limited by technical means of collective protection or procedures of work organisation.

The second directive, Directive 89/686/EEC – *Personal Protective Equipment* was adopted on December 21, 1989 by the European Council. It defines essential requirements which PPE must satisfy at the time of manufacture and before it is placed on the market: the general requirements applicable to all PPE; the additional requirements specific to certain types of PPE and also the additional requirements specific to particular risks. The aim of the directive is to ensure the free movement of PPE within the community market by completely harmonising the essential safety requirements to which it must conform (EU-OSHA website). Directive-89/686 proposed testing procedures, manufacturing and marketing of personal protective equipment, which as a consequence stopped the introduction of national standards for members of the EU. A large number of standards were developed by the European standardisation organisation (CEN) in order to enable manufacturers to test their products and guarantee compliance with those directives (Holmér, 2006; CEN: European Committee for Standardization, 2013).

In 1994, the European agency for safety and health at work (EU-OSHA) was established by Council Regulation (EC) No 2062/943 to contribute to improving working environments, as regards to the protection of the safety and health of workers and by implementing the actions designed to increase and disseminate knowledge likely to assist this improvement. In April 21, 2018, EU-OSHA repealed Directive 89/686/EEC and replaced it with a new regulation, (EU) 2016/425 *On Personal Protective Equipment* (EU-OSHA website). What is interesting is that this regulation applies to all sorts of PPE except, for example, for those used by the armed forces, certain private uses of PPE and the ones subject to other rules (e.g. PPE on seagoing vessels, motorcycle helmets, etc.).

In 1989, the working group WG4, under the technical committee TC162 of the European Committee for Standardization (CEN), developed a proposal for a standard for protective clothing against cold, called the EN 342 *Protective Clothing – Ensembles and Garment for Protection Against Cold* (BS EN 342:2004). Clothing ensembles were tested by a walking thermal manikin deriving the basic insulation value of the garment and ensembles. Later, the details of the manikin testing were excluded, and in 2003 ISO 15831 *Clothing – Physiological effects – Measurement of Thermal Insulation by Means of a Thermal Manikin* was created (ISO 15831:2004).

The general requirements are discussed by European standard BS EN 340 (BS EN 340:2003) which was later replaced by international standard ISO 13688-2013 (ISO 13688:2013).

CEN/TC162/WG8 developed a standard for protective gloves against cold in 1993 and revised it in 2004 and 2006 (BS EN 511 *Protective Clothing – Protective Gloves Against Cold*) (BS EN 511:2006). A complete glove is tested on a thermal model of a hand. In 2002, CEN published a standard for measuring sleeping bags on the basis of the German standard DIN EN 13537:2002-11 (DIN EN 13537:2002-11), which was revised in 2012. The standard specifies requirements for sleeping bags and methods for measurement, including measurement with a thermal manikin. CEN TC161/WG1 decided to investigate the option of revising EN 344-1-1993 (BS EN 344-1:1993), in terms of testing the cold protection properties of footwear (Holmér, 2006). This standard was replaced by *ISO 20344:2004* and updated in 2011 (*ISO 20344:2011*).

Along with ISO 20344, the other standards for safety, protective and occupational footwear were created in 2004. These are ISO 20345 *Personal Protective Equipment – Safety Footwear* (ISO 20345:2011), ISO 20346 *Personal Protective Equipment – Protective Footwear* (ISO 20346:2014), ISO 20347 *Personal Protective Equipment – Occupational Footwear* (ISO 20347:2012).

7.3 FURTHER DEVELOPMENT AND NECESSARY IMPROVEMENTS

International standards are continually reviewed in terms of new knowledge and requirements and thus provide the best available methods for the future. Future requirements for standards will have to consider the sustainable use of resources, additional requirements in occupational environments and energy efficiency (Parsons, 2013). However, there are courses that need to be addressed based on the observed

The Standardisation of Thermal Comfort

disadvantages in the current standards or the lack of some important methods for future development.

There is a problem involving the connection between numerous factors affecting humans in working environments and their application as an integrated and interconnected complex unit within any standard on thermal comfort and the scope of the ergonomics of the thermal environment. The applied ergonomist has to be able to simultaneously consider all relevant factors that affect the human occupant. These are factors affecting the health, comfort and performance of the occupants, as well as the environmental factors (Parsons, 1999).

Nowadays, national or international standards on thermal comfort assessment and associated issues move towards greater concern on climate change and the consequential impact on human well-being. In the future, extreme climate change is expected, forcing humans to prepare for harsher ambient conditions. This may have a significant impact on occupational health, heat and cold exposures, and raises concern about whether the current system of international standards is appropriate for future climatic influences. A shortcoming for ISO standards is the absence of the connection between meteorological data and applicable measurements or demands required by current standards. Future work on thermal comfort standardisation should provide a simple basis for predicting the global consequences of a changing climate and anticipating potential impacts on occupational health (Parsons, 2013).

Current international standards are not necessarily truly successful in being applicable to all people and all climatic conditions (Parsons, 2008). The future work on standardisation is aimed at covering the increasing need for extended application such as contact with hot, moderate and cold solid surfaces, quantities symbols and units, requirements for users with special needs, thermal environments in vehicles, working practices for cold environments, and the thermal performance of buildings (Parsons, 1999). One of the most prominent changes in the last few years is the psychology behind the standards, since the new idea is brought to light that the international standards should move from an application based upon results from massive trials and general considerations to more adaptive implementation based on individual needs (Parsons, 2002).

Alterations within the acceptable thermal conditions predicted to be satisfied by ISO 7730, include designing for a smaller number of people of different ages, sex and fitness engaged in a variety of activities. Existing thermal comfort standards base their results upon research carried out on a large human sample, but in reality most of the spaces are occupied by a small human sample. Designing spaces for satisfactory thermal comfort indices for a small number of people will lead to facing a larger number of individual differences. Since the metabolic rates variations can be as much as 20% for sex and age differences, and as much as 50 to 90% for weight and fitness differences. Fanger's equation for thermal comfort could be extended to a broader range of acceptable comfort temperatures. The current proposed version implicates the thermal comfort conditions satisfactory with less than 10% predicted percentage of dissatisfied (PPD) when considering individual differences (Ong, 1995). There is also the lack of thermal environment classification and the selection the appropriate category discussed by some authors. A proposed classification scheme, taking into account individual thermal sensitivity, task accuracy and thermal manipulability of the building, was developed by Lenzuni et al., which could represent the basis for the development of future normative documents (Lenzuni et al., 2009.).

The needs of people with disabilities, people with special health requirements, children and the elderly, as well as the thermal comfort differences due to gender, will have to be addressed (Kyoung An and Domina, 2015; Olesen et al., 2016). On a global scale, there are more than 1000 million people with disabilities (15% of the world's population). According to the World Health Organisation, one of the proposed inputs in the current action plan is to develop or revise legislation, policies, standards and regulatory mechanisms in order to raise the quality of life for people with disabilities (*WHO Global Disability Action Plan 2014–2021*, 2015). People with a physical disability have altered metabolic rates and consequently limited adaptive control to their environments, which could lead to rating otherwise acceptable environments as unacceptable. The study by Webb and Parsons (Webb and Parsons, 1998) and Hill et al. showed a lack of understanding for special needs and acceptable thermal environments for people with physical disabilities (Hill et al., 2000). Altered adaptive considerations, such as additional requirements on behavioural control, impact of prescribed drugs that affect the metabolic heat production, altered thermoregulatory capacity, reduced activity due to immobility and so on, should be taken into consideration when enhancing the international standards in order to better correspond to needs of people with disabilities (Webb and Parsons, 2000; Parsons, 2002). The International Standards Organisation already identified a need for creating a standard on thermal comfort requirements for people with a disability (Parsons, 2003). The only international standard addressing the needs of people with special requirements in moderate, cold and hot environments is ISO/TS 14415:2005. The special considerations taken into account for people with special requirements and disabilities include sensory impairment, paralysis, body shapes, impairment of the sweat secretion and vasomotor control, differences in metabolic rates and impaired physiological functions due to the heat stress exposures. The number of factors used to calculate the PMV thermal comfort index could be altered for the population of people with special requirements based on observed variations in metabolic variables. The creation of a new standard with the adaptive approach has proven to be a far better solution than adapting the existent standards on the subject (Parsons, 2000b; ISO 14415:2005). However, the modification of the above-mentioned standard is in progress, and the 2005 version is currently withdrawn (ISO website).

Since user-oriented application is a prime concern of international standards, their applicability and the scope of performance can be further extended due to new methods and energy use (Olesen et al., 2016). There is also a great need for standards to expand their operating field in order to cover heat stress due to unexpected weather anomalies and global warming. There is much work to be done in order to standardise and envision the requirements set upon protective clothing, health care and performance in different, often extreme, working environments, to prevent health risk and to form appropriate heat stress management systems (Parsons, 2015).

Humans are subjective beings. What seems to be satisfactory for one person is not necessarily satisfactory for another. Therefore, from this basic definition, one can conclude that thermal comfort is a state of mind rather than a state variable.

The Standardisation of Thermal Comfort

Although the work and applied physiology developed from 1940s onward, accompanied by the simultaneous development of equipment, mostly to avoid subjectivity during human testing and to avoid the life endangerment, the secondary application of findings weren't applied to final products until the 1980s. At least not on European soil. The requirements on clothing needed for specific usage were proposed as late as 1986. Although standards, which are voluntary, were created in 1986, but not until 1989 was there a regulation which was mandatory for Europe. Today, there are no standards setting the requirements for everyday clothing. Nor has there ever been one for the general public. Whether someone wants to know if their jacket has sufficient thermal protection, they have to rely on their own personal preferences. One can only see the criterion set on personal protective equipment. However, current standards for safety, protective and occupational footwear are not yet satisfactory and need to be revised.

All of the standards on protective footwear, such as EN 344-1:1993 (BS EN 344-1:1993), ISO 20344:2011 (ISO 20344:2011), ISO 20345 *Personal Protective Equipment – Safety Footwear* (ISO 20345:2011), ISO 20346 *Personal Protective Equipment – Protective Footwear* (ISO 20346:2014) and ISO 20347 *Personal Protective Equipment – Occupational Footwear* (ISO 20347:2012), classify footwear as cold-protective by a pass/fail test where the limits are set for an allowed 10°C temperature drop inside the footwear during 30 minutes at a temperature gradient of ~40°C. Over the years, the standards have been reviewed. However, the test on cold protection has got only minor improvement, mainly directed towards making the testing procedure simpler and less time-consuming. Concerns have been raised to demonstrate that those standard test methods are neither relevant nor valid for testing and classification of cold-protective footwear (Kuklane et al., 2009).

The present ISO standards do not measure insulation but classify footwear as cold-protective by a simple pass/fail test. For example, the same footwear that achieves good thermal comfort at 263.15 K (−10°C) when walking may be too cold for standing at the same temperature or walking at 248.15 K (−25°C), and too warm to be used at 283.15 K (+10°C). It is also the case that practically any professional footwear can pass the test. If insulation, environmental temperature and activity level (heat input to feet) is known, then it is possible to choose correct footwear with the required insulation for particular conditions. It is also possible to estimate the change in foot skin temperatures: by considering dynamic changes one can determine the recommended exposure time according to threshold limit values for cold sensation, strong cold sensation or pain sensation. The thermal foot method allows for an evaluation of the footwear as an entity and provides feedback to manufacturers on the footwear as a whole as well as on separate areas. The method also provides useful information to customers. The results can be used in prediction models to estimate the insulation need, exposure time or what happens under certain conditions, and provide recommendations for use accordingly (Kuklane, 2009). It is only a question of time as to when the thermal foot model will become the standardised method for requiring insulation values for footwear. There are also European standards which haven't yet been accepted on an international level for other protective accessories as equally important as any other protective item, such as BS EN 407:2004 *Protective Gloves Against Thermal Risks (Heat And/Or Fire)* (BS EN 407:2004).

In the last decade, the research on various aspects of protective clothing and accessories yielded some interesting findings. However, European standards such as BS EN 342 (BS EN 342:2004) and BS EN 511 (BS EN 511:2006), still missed relevant information on handwear for protection against cold and frostbite. Only basic requirements and testing procedures for protective gloves against cold are summarised in BS EN 511 in terms of classes. In the future, the experimental results from various studies should be incorporated into standards to create operating instructions. Zimmermann et al. worked towards initiating the development of standards and evaluation procedures for handwear systems (Zimmermann et al., 2008). Currently, the international standard on protective gloves is in its last stages of development (ISO/FDIS 21420) based on the European standard (BS EN 420 (2003) + A1; 2009, 2003).

Nowadays, although the wars are still present, new materials and clothing systems enhance standard development, or at least the testing procedures. There are several existing standards on testing procedures, which are used as supporting standards in the area of thermal comfort, since none of the thermal performance measurements performed on clothing go without testing the corresponding material properties such as air permeability, water penetration and thermal and water vapour resistance. These are ISO 5085-1 (ISO 5085-1:1989), ISO 5085-2 (ISO 5085-2:1990), ISO 811 (ISO 811:2018), ISO 11092 (ISO 11092:2014) and ISO 9237 (ISO 9237:1995).

Today, new materials are developed on an almost daily basis, one example being the different types of composite materials. The term composite materials is largely misunderstood. Even in the case of traditional isotropic homogeneous materials such as steel, copper, aluminium or plastics, there is little real understanding of why they are used for particular applications. Knowledge on the way that homogeneous materials react to the application of loads of varying types is not sufficient to predict, for example, the mechanical properties of advanced fibre composites. Nowadays, there are numerous initiatives by standards organisations worldwide to update existing methods and produce new standard test methods to satisfy the particular requirements of advanced fibre-reinforced plastic matrix composite materials. In the 1980s and 1990s, many organisations were involved in standards developments for composite material testing, especially the Composites Research Advisory Group by the UK's Ministry of Defence (CRAG), which attempted to define best practice over a range of test methods. The CRAG recommendations were proposed to the British Standards Institution and subsequently had a considerable effect on the development of new international standards. One of them is known as ISO 14126:1999, *Fibre-Reinforced Plastic Composites – Determination of Compressive Properties in the In-Plane Direction* (ISO 14126:1999). Crucial elements of this standard were also adopted within the ASTM D 3410/D 3410M-95, *Standard Test Method for Compressive Properties of Polymer Matrix Composite Materials With Unsupported Gauge Section by Shear Loading* (ASTM D 3410/D 3410M-95, 1997).

The new materials actuate the need for standards improvement and creation. There is a great example when tracking the development of a shape-memory alloy (SMA). The SMA components, employed as actuators, has enabled a number of adaptable aero-structural solutions. Unfortunately, there are currently no industry or government-accepted standardised test methods for SMA materials. The new specification and standardisation efforts are expected to deliver the first ever-regulatory agency-accepted material specification and test standards for SMA, when used as actuators and their transition to commercialisation and production for commercial and military aviation applications. The work is done to deliver a set of agreed-upon properties to be included in future material certification specifications as well as the associated experiments needed to obtain them in a consistent manner. The participation and input from a number of organisations and individuals is essential, including engineers and designers working in materials and processing development, application design, SMA component fabrication and testing at the material, component, and system level. In addition, a strong consensus among this diverse body of participants and the SMA research community at large is needed to advance standardisation concepts for universal adoption by the greater aerospace community and especially regulatory bodies (Hartl et al., 2015).

The research studies on clothing often requires a multidisciplinary approach involving many different areas, especially when measuring the insulation provided by the garments or the overall ensembles. Environmental conditions tend to change the performance of the clothing substantially. In addition, one has to bear in mind also the impact of fabrics and design on final thermal performance.

The layers of a clothing system can interact differently than expected upon their individual properties. With that in mind, the clothing system designers can face a challenge when ensuring the appropriate selection of the materials for each layer, and these layers interact (Morrissey and Rossi, 2013). When fabrics are worn as an ensemble, the total clothing insulation is more than the sum of the thermal resistance of each material layer. According to Havenith et al. for a two-layer clothing ensemble, more than 60% of the total insulation can be attributed to air layers between the skin and the fabric and between the layers of the fabric, and less than 40% can be attributed to the clothing layers themselves (Havenith et al., 1990).

There have been a number of studies of the insulation and the ventilation properties of clothing. Dry thermal insulation values have been determined for many types of clothing using thermal manikins. The large database of clothing insulation values for a number of different ensembles have been developed by McCullough et al. (McCullough and Jones, 1984; McCullough et al., 1985, 1989; McCullough and Hong, 1992) and Olesen et al. (Olesen, 1982; Olesen and Nielsen, 1983), which were later incorporated into ISO 9920 (ISO 9920:2008). The use of clothing tables is not unambiguous. McCullough et al. showed that even trained textile students experience difficulties in identifying clothing items from those listed in the tables, the result being a standard deviation of 23% in the intrinsic insulation (McCullough et al., 1985). A study from Lotens showed interesting findings (Lotens, 1989). Since enclosed air layers constitute a considerable part of total insulation, it is anticipated that a change in the fit of clothing or in posture affects the insulation. Sitting posture showed a slight increase in insulation for light clothing and a slight decrease for heavy clothing in comparison to standing. Both wind and body motion may affect insulation through the induction of air motion. When outside air penetrates the clothing, either via closures or through the material, the effect is called ventilation. When there is no air exchange, but rather an increased circulation underneath the clothing or at the surface of the clothing, it is called convection. Usually, ventilation and convection will accompany one another, Figure 7.1.

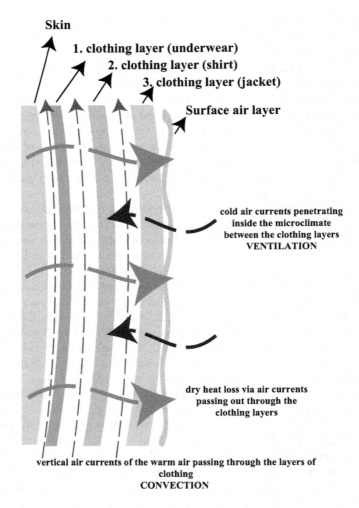

FIGURE 7.1 The air currents through clothing layers.

After testing, Lotens has shown that the intrinsic clothing insulation is not a specific clothing constant but depends instead on air convection. The concept of intrinsic insulation is meaningful only for a standard environment, for example, indoors. He has also found the effects of motion and wind are so strong that for many outdoor conditions the insulation is less than half the table value provided by McCullough et al. (Lotens, 1989).

The clothing insulation value is actually lower during movement. In 1992, McCullough and Hong investigated the changes in insulation values of typical indoor clothing between a stationary and walking manikin (no wind) (McCullough and Hong, 1992).

Havenith et al. (1990) proved that wind up to 4 m/s has a small effect on the intrinsic clothing insulation because it doesn't penetrate materials and thus does not disturb the entrapped air layers (Havenith et al., 1990). However, the total clothing insulation reduces when walking (Havenith et al., 1990; Holmér et al., 1992, 2002; Wagner and Dorawa, 2016) but microclimate is also affected by ventilation according to Lim et al. (Lim et al., 2015). MacRea et al. reported that outerwear and additional air spaces appeared to dominate the overall thermal properties of clothing ensembles (MacRea et al., 2011).

Although earlier studies from Goldman and his colleagues suggested that a walking manikin was unnecessary since these pumping coefficients could be generated simply by using three different wind speeds. During this study, the first walking manikin from copper was produced to evaluate it. He reported that the subsequent studies suggested that a walking manikin was unnecessary since these pumping coefficients could be generated simply by using three different wind speeds (Goldman, 2006). However, the research conducted by Ingvar Holmér proved that the combined effects of body movements and wind considerably enhance heat loss from the human body, since the resultant insulation provided by clothing is reduced (Holmér et al., 1992).

Today there are also several studies on convection air currents inside the clothing garments. Clothing entraps air next to the skin in the weave of the cloth, thereby increasing the air thickness adjacent to the skin while simultaneously decreasing the flow of convection air currents (Guyton and Hall, 2016). E. A. McCullough and associates found clothing to reduce conduction losses by trapping still air within the fabric structures and between the garment layers. It resists convective losses by preventing convection currents forming next to the body and by providing a barrier against air currents in the environment (McCullough et al., 1985). The type and the specific design of the clothing garment together with the fabric selection will determine the fit to the body, thus creating nonhomogeneous air layers of a different thickness under the clothing. Thermal insulation increases linearly with an air gap as long no convection is present (Lee et al., 2007; Mert et al., 2015; Spencer-Smith, 1977; Li et al., 2013; Zhang and Li, 2011). But complex geometry of those layers makes it hard to quantify the volume of the microclimatic air.

Havenith refers to the *'microclimate heat pipe'* as heat being transported by vapour and released further by condensation (Havenith et al., 2008). Heat is released to the outer clothing layers without evaporating into the environment. When ambient temperature values drop below the skin temperature, the microclimate heat pipe effect will occur, which transfers latent heat from the skin to clothing (from where it is lost to the environment by increased radiation and convection) without losing moisture (weight) from the clothing. One has to distinguish the impact of the natural convection of the air inside the microclimatic layers from forced convection due to compression of the garments while performing physical activities.

When the air volume exceeds a certain limit, it becomes an ineffective insulator (Lee et al., 2007) as forced convection due to body movements create a *bellows effect*, thus preventing the linear increase of total dry heat loss (Cena and Clark, 1981; Havenith, 2005; Karwowski, 2006; Rainford and Gradwell, 2006; Clark and de Calcina-Goff, 2009; Mert et al., 2015).

T. S. Mayor et al. have proven that microclimate thickness plays an important role in the onset of natural convection. According to this study, 91 to 92% of the heat is transferred by convection with low-emissivity fabrics and microclimates above 25 mm of width (Mayor et al., 2015). The natural convection will only occur if there is a difference between the temperature of the skin and the surrounding air, while humans are at rest and when there is no substantial ambient air movement. During body movements and exercise, heat is transferred from the body by forced convection. During rest, natural convection is the dominant heat exchange mechanism, but according to Zhang et al. this is the case only when air gap thickness is larger than 15 mm (Zhang et al., 2011). Most of the experiments to prove the impact of natural convection are made in still ambient air below 1 m/s. When the body is still, the natural convection or *chimney effect* occurs, and the microclimatic air ventilates through upper clothing openings. During the recovery period, after exercise, natural convection is more pronounced and tight neck openings cause uncomfortable sensations to the wearer (Zhang et al., 2012). According to N. Ismail et al., with wind velocities higher than 2 m/s, the natural convection effect can be neglected when estimating clothing ventilation (Ismail et al., 2014).

The multidisciplinary approach is more than evident when summarising all of the impacts on the overall thermal insulation of the clothing. There is another example of broadening the range of methods found in acquiring the microclimatic volume under clothing. For example, Daanen et al. found 3D scanning to be the most accurate and reproducible method for air volume quantification under clothing (Daanen et al., 2002). According to Lee et al. (2007), Zhang and Li (2011), Psikuta et al. (2012), Daanen and Psikuta (2018), Carfagnia et al. (2017), 3D scanning technology made it possible to precisely quantify the microclimatic layers.

In the past, McCullogh and Hong quantified the thickness of air and fabric layers by subtracting nude body radius from the outer clothing radius, thus calculating the thickness of the overall microclimatic layer (McCullough and Hong, 1994). The volume was ultimately approximated by cylinders representing different parts of the body, although this is merely a rough overview. Lee et al. quantified volume of the microclimatic layers by using 3D phase-shifting moiré topography (Lee et al., 2007). However, this specific method did not provide sufficient data on precise calculations nor thickness nor volume of the air gaps because of the considerable variations due to body motion and clothing draping or folding. This technique was time efficient and accurate, but it was not reported if this technique could be used on humans and how the results could be compared to the traditional technique of microclimate volume measurement (Daanen et al., 2002). Zhang and Li came to a similar conclusion (Zhang and Li, 2011). But none of the studies mentioned above investigated three-layered ensembles and the impact of the outer covering garments.

The developing body ventilation systems (BVS) for specific use, such as external active or passive cooling systems under chemical protective suits (Vernieuw et al., 2007; Xu and Gonzalez, 2011) or personal heating garments (PHGs) and heated technology for immersion suits (Wang et al., 2010) also fosters the development of standards. Currently, there are only two standards on measuring the performance of personal cooling systems with sweating heated thermal manikins and physiological testing, ASTM F 2300-10 (ASTM F2300-10, 2016) and ASTM F 2371-16 (ASTM F2371-16, 2016), and no standards on evaluating the performances of personal heating garments (Wang et al., 2010). Only standards on test methods and performance requirements for protective clothing for firefighters have been revised and updated (ISO 13506-1:2017; ISO 13506-2:2017; ISO 15384:2018).

It is evident that although clothing has been worn for tens of thousands of years, and although it is a basic human need together with food and shelter, the research on many different parameters of clothing are yet to continue in order to fully understand the specific requirements set upon clothing. This will inevitably foster the development of new standards.

Heat and cold stress can also occur in both outdoor occupations and also indoor settings, such as manufacturing and food processing, and in work with minimal physical activity. When the thermal balance is disturbed, thermal stress presents a multitude of potential issues that may reduce operational capacity or increase acute and chronic health risks for the worker. Overall, while there is a large body of knowledge concerning the physiological effects of thermal stress, work into direct occupational applications such as developing physical employment standards can be considered to be in its infancy. Physical employment standards are largely nonexistent in most cold-related occupations (Cheung et al., 2016).

Most of the standards currently on the market are aimed at protecting the worker's health from extreme environments and to specify conditions for comfort or to predict the degree of discomfort. A new initiative has been launched to develop a series of international standards on the effects of the physical environment on human performance, among others heat since there is still a lack of international standards on the subject. A fine balance has to be made between the cost of making new standards and benefits provided due to their application. The goal is to specify appropriate physical environments that will provide optimal human performance and maximum working productivity (Parsons, 2018).

REFERENCES

A brief history of thermal comfort: From effective temperature to adaptive thermal comfort, Chapter 2. 2015. In: *Indoor Thermal Comfort Perception*, ed. K. Fabbri. Switzerland: Springer International Publishing.

ANSI/ASHRAE Standard 55-1966 Thermal Environmental Conditions for Human Occupancy. 1966. Atlanta, GA: American Society of Heating, Refrigerating and Air Conditioning Engineers Inc. (ASHRAE).

ANSI/ASHRAE Standard 55-1992 Thermal Environmental Conditions for Human Occupancy. 1992. Atlanta, GA: American Society of Heating, Refrigerating and Air Conditioning Engineers Inc. (ASHRAE).

ANSI/ASHRAE Standard 55-2004 Thermal Environmental Conditions for Human Occupancy. 2004. Atlanta, GA: American Society of Heating, Refrigerating and Air Conditioning Engineers Inc. (ASHRAE).

ANSI/ASHRAE Standard 55-2010 Thermal Environmental Conditions for Human Occupancy. 2010. Atlanta, GA: American Society of Heating, Refrigerating and Air Conditioning Engineers Inc. (ASHRAE).

ANSI/ASHRAE Standard 55-2017 Thermal Environmental Conditions for Human Occupancy. 2017. Atlanta, GA: American Society of Heating, Refrigerating and Air Conditioning Engineers Inc. (ASHRAE), ISSN: 1041-2336.

ASHRAE Handbook Fundamentals (SI ed.). 2017. Atlanta, GA: American Society of Heating, Refrigerating, and Air Conditioning Engineers Inc., ISBN: 1931862702.

ASTM. The history of ASTM international. https://www.astm.org/ABOUT/history_book.html (accessed on April 1, 2013).

ASTM D 3410/D 3410M-95 Standard Test Method for Compressive Properties of Polymer Matrix Composite Materials with Unsupported Gage Section by Shear Loading. 1997. Annual Book of ASTM Standards, Volume: 15.03. Philadelphia, PA: ASTM International, 116–131.

ASTM F 2371-16 Standard Test Method for Measuring the Heat Removal Rate of Personal Cooling Systems Using a Sweating Heated Manikin. 2016. Book of Standards Volume: 11.03. Philadelphia, PA: ASTM International.

ASTM F2300-10 Standard Test Method for Measuring the Performance of Personal Cooling Systems Using Physiological Testing. 2016. Philadelphia, PA: ASTM International.

Bai, L., S. Carlucci and L. Yang. 2017. Critical analysis of the application of the adaptive thermal comfort models proposed by standards. In: *OB-17 Symposium on Occupant Behaviour and Adaptive Thermal Comfort.* Denmark: Technical University of Denmark.

Bell, J. 2004. *Indoor Environments: Design, Productivity and Health.* Report Number 2001-005-B final report, Volume 2, literature database, Icon.Net Pty Ltd.

Bell, P. A. 1980. Effects of heat, noise and provocation on retaliatory evaluative behaviour. *The Journal of Social Psychology* 40: 97–100.

Bell, P. A. 1981. Physiological, comfort, performance, and social effects of heat stress. *Journal of Social Issues* 37 (1): 71–94.

Bell, P. A. and T. C. Greene. 1982. Thermal stress: Physiological, comfort, performance and social effects of hot and cold environments. In: *Environmental Stress*, ed. G. W. Evans. London, UK: Cambridge University Press.

Brager, G. S. and R. de Dear. 2000. A standard for natural ventilation. *ASHRAE Journal* 42(10): 21–27.

Brager, G. S., G. Paliaga, R. De Dear, B. W. Olesen, J. Wen, F. Nicol and M. Humphreys. 2004. Operable windows, personal control, and occupant comfort. *ASHRAE Transactions* 110(2): 17–35.

Bröde, P., K. Błażejczyk, D. Fiala, G. Havenith, I. Holmér, G. Jendritzky, K. Kuklane and B. Kampmann. 2013. The Universal Thermal Climate Index (UTCI) compared to ergonomics standards for assessing the thermal environment. *Industrial Health* 51: 16–24.

BS EN 340:2003 Protective Clothing. General Requirements. 2003. BSI.

BS EN 342:2004 Protective Clothing – Ensembles and Garment for Protection Against Cold. 2004. BSI.

BS EN 344-1:1993 Safety, Protective and Occupational Footwear for Professional Use. Requirements and Test Methods. 1993. BSI.

BS EN 407:2004 Protective Gloves Against Thermal Risks (Heat and/or Fire). 2004. BSI.

BS EN 420 (2003) + A1 (2009) Protective Gloves. General Requirements and Test Methods. BSI. 2003.

BS EN 511:2006 Protective Clothing – Protective Gloves Against Cold. 2006. BSI.

Carfagnia, M., R. Furferia, L. Governia, M. Servia, F. Uccheddua, Y. Volpea and K. Mcgreevy. 2017. Fast and low cost acquisition and reconstruction system for human hand-wrist-arm anatomy. *Procedia Manufacturing* 11: 1600–1608.

CEN: European Committee for Standardization, Annual report. 2013. https://www.cen.eu/news/brochures/brochures/AR2013_CEN_EN_final.pdf (accessed on May 13, 2018).

Cheung, S. S., J. K. W. Lee, and J. Oksa. 2016. Thermal stress, human performance, and physical employment standards. *Applied Physiology, Nutrition, and Metabolism* 41: 148–164.

Clark, R. P. and M. L. de Calcina-Goff. 2009. Some aspects of the airborne transmission of infection. *Journal of the Royal Society Interface* 6(6): 767–782.

Daanen, H. A. M. and A. Psikuta. 2018. 3D body scanning, Chapter 10. In: *Automation in Garment Manufacturing, A volume in The Textile Institute Book Series*, 1st edition, ed. R. Nayak and R. Padhye. Langford Lane, Kidlington, UK: Elsevier Ltd., ISBN: 978-0-08-101211-6.

Daanen, H., K. Hatcher and G. Havenith. 2002. Determination of clothing microclimate volume. In: *Proceedings of the 10th International Conference on Environmental Ergonomics*, Fukuoka, Japan, Environmental Ergonomics X, 665–668.

de Dear R. J. and G. S. Brager. 1998a. Developing an adaptive model of thermal comfort and preference. *ASHRAE Technical Data Bulletin: Field Studies of Thermal Comfort and Adaptation* 14(1): 27–49.

de Dear, R. J. and G. S. Brager. 1998b. Towards an adaptive model of thermal comfort and preference. *ASHRAE Transactions* 104(1): 145–167.

de Dear, R. J. and G. S. Brager. 2002. Thermal comfort in naturally ventilated buildings: Revisions to ASHRAE Standard 55. *Energy and Buildings* 34: 549–561

de Dear, R. J., G. S. Brager and D. Cooper. 1997. *Developing an Adaptive Model of Thermal Comfort and Preference*, Final Report ASHRAE RP-884, ASHRAE.

de Dear, R. J., T. Akimoto, E. A. Arens, G. Brager, C. Candido, K. W. D. Cheong, B. Li, N. Nishihara, S. C. Sekhar, S. Tanabe, J. Toftum, H. Zhang and Y. Zhu. 2013. Progress in thermal comfort research over the last twenty years, *Indoor Air* 23: 442–461.

DIN EN 13537:2002-11 Requirements for Sleeping Bags (ger. Anforderungen an Schlafsäcke). 2002. DIN.

Endrusick, T. L., W. R. Santee, D. A. DiRaimo, L. A. Blanchard and R. G. Gonzalez. 1992. Physiological responses while wearing protective footwear in a cold–wet environment. In: *Performance of Protective Clothing, Volume 4 (STP 1133)*, ed. J. McBriarty and N. Henry. Philadelphia, pa: American Society of Testing and Materials, 544–556.

Ersting's Aviation Medicine, 4th edition. 2006. ed. D. J. Rainford and D. P. Gradwell. London, UK: Hodder Education, ISBN: 978-0-340-81319-5.

Fanger, P. O. 1970. *Thermal Comfort, Analysis and Applications in Environmental Engineering*. Copenhagen, Denmark: Danish Technical Press.

Fanger, P. O. 1973a. Assessment of man's thermal comfort in practice. *British Journal of Industrial Medicine* 30: 313–324.

Fanger, P. O. 1973b. *Thermal Comfort: Analysis and Applications in Environmental Engineering*. New York, NY: McGraw-Hill Book Company, ISBN: 0-07-019915-9.

Fanger, P. O. 1982. *Thermal Comfort*. Malabar, FL: Robert E. Krieger Publishing Company.

Fiala, D., G. Havenith, P. Bröde, B. Kampmann and G. Jendritzky. 2012. UTCI-Fiala multi-node model of human heat transfer and temperature regulation. *International Journal of Biometeorology* 56: 429–441.

Fiala, D., K. J. Lomas and M. Stohrer. 1999. A computer model of human thermoregulation for a wide range of environmental conditions: The passive system. *Journal of Applied Physiology* 87(6): 1957–1999.

Friendship Among Equals, Recollection from ISO'S First Fifty Years. 1997. ISO, 1–80, ISBN 92-67-10260-5.

Gagge, A. P. 1936. The linearity criterion as applied to partitional calorimetry. *American Journal of Physiology* 116: 656–668.

Gagge, A. P. 1937. A new physiological variable associated with sensible and insensible perspiration. *The American Journal of Physiology* 120 (2): 277–287.

Gagge, A. P. and Y. Nishi. 1976. Physical indices of the thermal environment. *ASHRAE Journal* 18 (1): 47–52.

Gagge, A. P., A. P. Fobelets and L. G. Berglund. 1986. A standard predictive index of human response to the thermal environment. *ASHRAE Transactions* 92(2B): 709–731.

Gagge, A. P., J. A. J. Stolwijk and Y. Nishi. 1971. An effective temperature scale based on a simple model of human physiological regulatory response. *ASHRAE Transactions* 77(1): 247–262.

Givoni, B. and R. F. Goldman. 1972. Predicting rectal temperature response to work, environment, and clothing. *Journal of Applied Physiology* 32(6): 812–822.

Givoni, B. and R. F. Goldman. 1973. Predicting heart rate response to work, environment and clothing. *Journal of Applied Physiology* 34 (2): 201–204.

Goldman, R. F. 2006. Thermal manikins, their origins and role. In: *Thermal Manikins and Modelling, 6th International Thermal Manikin and Modelling Meeting*, ed. J. Fan. Hong Kong: The Hong Kong Polytechnic University, 3–18, ISBN: 962-367-534-8.

Gonzales, R. R., Y. Nishi and A. P. Gagge. 1974. Experimental evaluation of standard effective temperature a new biometeorological index of man's thermal discomfort. *International Journal of Biometeorology* 18(1): 1–15.

Guyton, A. C. and J. E. Hall. 2016. *Guyton and Hall Textbook of Medical Physiology*, 13th edition, Philadelphia, PA: Elsevier Inc., ISBN: 978-1-4557-7005-2.

Hartl, D. J., J. H. Mabe, O. Benafan, A. Coda, B. Conduit, R. Padan and B. Van Doren. 2015. Standardization of shape memory alloy test methods toward certification of aerospace applications. *Smart Materials and Structures* 24. doi:10.1088/0964-1726/24/8/082001.

Havenith G., R. Heus and W. A. Lotens. 1990. Resultant clothing insulation: A function of body movement, posture, wind, clothing fit and ensemble thickness. *Ergonomics* 33(1): 67–84.

Havenith, G. 2005. Clothing heat exchange models for research and application. In: *Proceedings of 11th International Conference on Environmental Ergonomics*, ed. I. Holmér, K. Kuklane and C. Gao. Ystad, Sweden: Thermal Environment Laboratory, EAT, Department of Design Sciences, Lund University, 66–73, ISSN: 1650-9773.

Havenith, G., M. G. M. Richards, X. Wang, P. Bröde, V. Candas, E. den Hartog, I. Holmér, K. Kuklane, H. Meinander and W. Nocker. 2008. Apparent latent heat of evaporation from clothing: Attenuation and 'heat pipe' effects. *Journal of the Applied Physiology*, 104: 142–149.

Hill, L. D., L. H. Webb and K. C. Parsons. 2000. Carers' views of the thermal comfort requirements of people with physical disabilities. *Proceedings of the Human Factors and Ergonomics Society Annual Meeting* 44(28): 716–719.

Holmér, I. 2006. Use of thermal manikins in international standards. In: *Thermal Manikins and Modelling, 6th International Thermal Manikin and Modelling Meeting*, ed. J. Fan. Hong Kong: The Hong Kong Polytechnic University, 3–18, ISBN: 962-367-534-8.

Holmér, I., C. E. Gavhed, S. Grahn and H. O. Nilsson. 1992. Effects of wind and body movements on clothing insulation – Measurement with a moveable thermal manikin. In: *Proceedings of the 5th international conference on environmental ergonomics*, ed. W. A. Lotens and G. Havenith. Soesterberg, The Netherlands: TNO-Institute of Perception, 66–67, ISBN 90-6743-227-X.

Holmér, I., H. O. Nilsson and H. Anttonen. 2002. Prediction of wind effects on cold protective clothing. In: *RTO HFM Symposium on Blowing Hot and Cold: Protecting Against Climatic Extremes*. Neuilly-sur-Seine Cedex, France: RTO/NATO, ISBN 92-837-1082-7.

Huizenga, C., H. Zhang and E. A. Arens. 2001. A model of human physiology and comfort for assessing complex thermal environments. *Building and Environment* 36: 691–699.

Humphreys M. and J. F. Nicol. 1998. Understanding the adaptive approach to thermal comfort. *ASHRAE Technical Data Bulletin: Field Studies of Thermal Comfort and Adaptation* 14(1): 1–14.

IEC. IEC history. http://www.iec.ch/about/history/ (accessed on August 20, 2018).

INSTA 352 Protective Clothing – Clothing Against Cold – Classification, Marking and Instruction. 1986. INSTA-CERT.

The Standardisation of Thermal Comfort

INSTA 353 Protective Clothing – Clothing Against Cold – Thermal Manikin for Testing of Thermal Insulation. 1986. INSTA-CERT.

INSTA 354 Protective Clothing – Clothing Against Cold – Testing of the Thermal Insulation of Garment Piece. 1986. INSTA-CERT.

International Encyclopaedia of Ergonomics and Human Factors, Volume 1, 2nd edition. 2006. ed. W. Karwowski. Baco Raton, FL: CRC Press, Taylor & Francis Group, ISBN: 978-0-415-30430-6.

Ismail, N., N. Ghaddar and K. Ghali. 2014. Predicting segmental and overall ventilation of ensembles using an integrated bioheat and clothed cylinder ventilation models. *Textile Research Journal* 84 (20): 2198–2213.

ISO. ISO/TC 159 ergonomics. https://www.iso.org/committee/53348.html (accessed on August 9, 2018).

ISO 11092:2014 Physiological Effects – Measurement of Thermal and Water-Vapour Resistance under Steady-State Conditions (Sweating Guarded-Hotplate Test). 2014. ISO-International Organization for Standardization.

ISO 13506-1:2017 Protective Clothing Against Heat And Flame – Part 1: Test Method for Complete Garments – Measurement of Transferred Energy Using an Instrumented Manikin. 2017. ISO-International Organization for Standardization.

ISO 13506-2:2017 Protective Clothing Against Heat and Flame – Part 2: Skin Burn Injury Prediction – Calculation Requirements and Test Cases. 2017. ISO-International Organization for Standardization.

ISO 13688:2013 Protective Clothing-General Requirements. 2013. ISO-International Organization for Standardization.

ISO 14126:1999 Fibre-reinforced Plastic Composites – Determination of Compressive Properties in the In-plane Direction. ISO-International Organisation for Standardization.

ISO 14415:2005 Ergonomics of the Thermal Environment – Application of International Standards to People with Special Requirements. 2005. ISO-International Organization for Standardization.

ISO 15384:2018 Protective Clothing for Firefighters – Laboratory Test Methods and Performance Requirements for Wildland Firefighting Clothing. 2018. ISO-International Organization for Standardization.

ISO 15831:2004 Clothing – Physiological Effects – Measurement of Thermal Insulation by Means of a Thermal Manikin. 2004. ISO-International Organization for Standardization.

ISO 20344:2004 Personal Protective Equipment. Test Methods for Footwear. 2004. ISO-International Organization for Standardization.

ISO 20344:2011 Personal Protective Equipment – Test Methods for Footwear. 2011. ISO-International Organization for Standardization.

ISO 20345:2011 Personal Protective Equipment – Safety Footwear. 2011. ISO-International Organization for Standardization.

ISO 20346:2014 Personal Protective Equipment – Protective Footwear. 2014. ISO-International Organization for Standardization.

ISO 20347:2012 Personal Protective Equipment – Occupational Footwear. 2012. ISO-International Organization for Standardization.

ISO 5085-1:1989 Textiles – Determination of Thermal Resistance – Part 1: Low Thermal Resistance. 1989. ISO-International Organization for Standardization.

ISO 5085-2:1990 Textiles – Determination of Thermal Resistance – Part 2: High Thermal Resistance. 1990. ISO-International Organization for Standardization.

ISO 7243:2017 Ergonomics of the Thermal Environment – Assessment of Heat Stress Using the WBGT (Wet Bulb Globe Temperature) Index. 2017. ISO-International Organization for Standardization (BS EN 27243:2017).

ISO 7730:1994 Ergonomics of the Thermal Environment – Analytical Determination and Interpretation of Thermal Comfort Using Calculation of the PMV and PPD Indices and Local Thermal Comfort Criteria. 1994. ISO-International Organization for Standardization.

ISO 7730:2005 Ergonomics of the Thermal Environment – Analytical Determination and Interpretation of Thermal Comfort Using Calculation of the PMV and PPD Indices and Local Thermal Comfort Criteria. 2005. ISO-International Organization for Standardization.

ISO 811:2018 Textiles – Determination of Resistance to Water Penetration – Hydrostatic Pressure Test. 2018. ISO-International Organization for Standardization.

ISO 9237:1995 Textiles – Determination of the Permeability of Fabrics to Air. 1995. ISO-International Organization for Standardization.

ISO 9920:2008 Ergonomics of the Thermal Environment – Estimation of Thermal Insulation and Water-Vapour Resistance of a Clothing Ensemble. 2008. ISO-International Organization for Standardization.

ISO/FDIS 21420 Protective Gloves – General Requirements and Test Method. ISO-International Organisation for Standardization.

Janssen, J. E. 1999. The history of ventilation and temperature control. *ASHARE Journal* 47–52.

John, B..Pierce Laboratory. About us. http://jbpierce.org/about-us/ (accessed on April 16, 2015).

Jüptner, H. 1984. ISO standards in the field of ergonomics: A report of ISO/TC 159 activities. *Applied Ergonomics* 15(3): 211–213.

Kontes, G. D., G. I. Giannakis, P. Horn, S. Steiger and D. V. Rovas. 2017. Using thermostats for indoor climate control in office buildings: The effect on thermal comfort. *Energies* 10: 1368.

Kuklane, K. 2009. Protection of feet in cold exposure. *Industrial Health* 47: 242–253.

Kuklane, K., S. Ueno, S.-I. Sawada and I. Holmér. 2009. Testing cold protection according to EN ISO 20344: Is there any professional footwear that does not pass? *Annals of Occupational Hygiene* 53(1): 63–68.

Kyoung An, S. and T. Domina. 2015. Thermal comfort difference on gender under military garment system using thermal manikin. *AATCC Journal of Research* 2(3): 1–5.

Lee, Y., K. Hong and S.-A. Hong. 2007. 3D quantification of microclimate volume in layered clothing for the prediction of clothing insulation. *Applied Ergonomics* 38: 349–355.

Lenzuni, P., D. Freda and M. Del Gaudio. 2009. Classification of thermal environments for comfort assessment. *Annals of Occupational Hygiene* 53(4): 325–332.

Li, J., Z. Zhang and Y. Wang. 2013. The relationship between air gap sizes and clothing heat transfer performance. *The Journal of the Textile Institute* 104(12): 1327–1336.

Lim, J., H. Choi, E. K. Roh, H. Yoo and E. Kim. 2015. Assessment of airflow and microclimate for the running wear jacket with slits using CFD simulation. *Fashion and Textiles* 2(1). doi:10.1186/s40691-014-0025-2.

Lotens, W. A. 1989. The actual insulation of multilayer clothing. *Scandinavian Journal of Work, Environment & Health* 15(1): 66–75.

MacRea, B. A., R. M. Laing and C. A. Wilson. 2011. Importance of air spaces when comparing fabric thermal resistance. *Textile Research Journal* 81(19): 1962–1965.

Mayor, T. S., S. Couto, A. Psikuta and R. M. Rossi. 2015. Advanced modelling of the transport phenomena across horizontal clothing microclimates with natural convection. *International Journal of Biometeorology* 59: 1875–1889.

McCullough, E. A. and B. W. Jones. 1984. *A Comprehensive Database for Estimating Clothing Insulation*. Kansas, KS: Institute for Environmental Research, Kansas State University, IER Technical Report 84–01.

McCullough, E. A., B. W. Jones and J. Huck. 1985. A comprehensive database for estimating clothing insulation. *ASHRAE Transactions* 91 (2A): 29–47.

McCullough, E. A., B. W. Jones and T. Tamura. 1989. A database for determining the evaporative resistance of clothing. *ASHRAE Transactions* 95(2): 316–328.

McCullough, E. A. and S. Hong. 1992. A database for determining the effect of walking on clothing insulation. In: *Proceedings of the Fifth International Conference on Environmental Ergonomics*, ed. W. A. Lotens and G. Havenith. Maastricht, The Netherlands.

McCullough, E. A. and S. Hong. 1994. A data base for determining the decrease in clothing insulation due to body motion. *ASHRAE Transactions* 100: 765–775.

Mert, E., A. Psikuta, M.-A. Bueno and R. M. Rossi. 2015. Effect of heterogenous and homogenous air gaps on dry heat loss through the garment. *International Journal of Biometeorology* 59: 1701–1710.

Military Quantitative Physiology: Problems and Concepts in Military Operational Medicine. 2012. ed. K. E. Friedl and W. R. Santee. USA: Office of the surgeon general, ISBN 978-0-1609108-7-6.

Morikawa, M. and J. Morrison. 2004. *Who Develops ISO Standards? A Survey of Participation in ISO's International Standards Development Processes*. California, USA: Pacific Institute for Studies in Development, Environment, and Security. https://pacinst.org/reports/iso_participation/iso_participation_study.pdf (accessed August 18, 2018).

Morrissey, M. P. and R. M. Rossi. 2013. Clothing systems for outdoor activities. *Textile Progress* 45(2–3): 145–181.

Nicol, J. F. and Humphreys, M. A. 1972. Thermal comfort as part of a self regulating system. In: *Proceedings of CIB Commission W45 Symposium – Thermal Comfort and Moderate Heat Stress*, ed. Langdon, Humphreys and Nicol, Building Research Establishment Report. London, UK: Her Majesty's Stationery Office (HMSO), 263–274.

Nishi Y. and A. P. Gagge. 1970a. Mean skin temperature for man weighted by regional heat transfer coefficients. *Federation Proceedings* 29 (2): 852.

Nishi Y. and A. P. Gagge. 1970b. Direct evaluation of convective heat transfer coefficient by naphthalene sublimation. *Journal of Applied Physiology* 29 (6): 830–838.

Nishi, Y. and A. P. Gagge. 1971. Humid operative temperature: A biophysical index of thermal sensation and discomfort. *Journal of Physiology (Paris)* 63(3): 365–368.

Nishi Y. and A. P. Gagge. 1973. Moisture permeation of clothing: A factor governing thermal equilibrium and comfort. *ASHRAE Transactions* 76(1): 365–368.

Nishi Y. and A. P. Gagge. 1974. A psychrometric chart for graphical prediction of comfort and heat tolerance. *ASHRAE Transactions* 80(2): 115–130.

Nishi Y. and A. P. Gagge. 1977. Effective temperature scale useful for hypo- and hyperbaric environments. *Aviation, Space, and Environmental Medicine (ASEM)* 48(2): 97–107.

Olesen, B. W. 1982. Thermal comfort. In: *Brüel and Kjaer Technical Review to Advance Techniques in Acoustical, Electrical and Mechanical Measurements*, number 2. Copenhagen, Denmark: Brüel and Kjaer.

Olesen, B. W. and R. Nielsen. 1983. *Thermal Insulation of Clothing Measured on a Moveable Thermal Manikin and on Human Subjects*. Lyngby, Denmark: Technical University of Denmark.

Olesen, B. W., F. R. d'Ambrosio Alfano, K. Parsons and B. I. Palella. 2016. The history of international standardization for the ergonomics of the thermal environment. In: *The book of proceedings of the 9th Windsor Conference: Making Comfort Relevant*, ed. L. Brotas, S. Roaf, F. Nicoland and M. Humphreys. Windsor, UK: Network for Comfort and Energy Use in Buildings (NCEUB), 15–38, ISBN: 978-0-9928957-3-0.

Ong, B. L. 1995. Designing for the individual: A radical reading of ISO 7730, Chapter 7. In: *Standards for Thermal Comfort: Indoor Air Temperature Standards for the 21st Century*, ed. F. Nicol, M. Humphreys, O. Sykes and S. Roaf. Milton Park, Abingdon, UK: Taylor & Francis, ISBN: O 419 20420 2.

OSHA. Safety and health legislation. https://osha.europa.eu/hr/legislation/ (accessed on April 20, 2017).

Pandolf, K. B., B. Givoni and R. F. Goldman. 1977. Predicting energy expenditure with loads while standing or walking very slowly. *Journal of Applied Physiology* 43(4), 577–581.

Parsons, K. 1995a. Ergonomics and international standards: History, organizational structure and method of development. *Applied Ergonomics* 26(4): 249–258.

Parsons, K. 1999. International standards for the assessment of the risk of thermal strain on clothed workers in hot environments. *Annals of Occupational Hygiene* 43(5): 297–308.

Parsons, K. 2001. Introduction to thermal comfort standards. http://www.utci.org/cost/publications/ISO%20Standards%20Ken%20Parsons.pdf (accessed on August 4, 2018).

Parsons, K. 2008. Industrial health for all: Appropriate physical environments, inclusive design, and standards that are truly international. *Industrial Health* 46: 195–197.

Parsons, K. 2013. Occupational health impacts of climate change: Current and Future ISO standards for the assessment of heat stress. *Industrial Health* 51: 86–100.

Parsons, K. 2015. ISO standards for the assessment of Heat stress. In: *Proceedings 19th Triennial Congress of the International Ergonomics Association (IEA)*, ed. G. Lindgaard and D. Moore. Melbourne, Australia: International Ergonomics Association.

Parsons, K. 2018. ISO standards on physical environments for worker performance and productivity. *Industrial Health* 56(2): 93–95.

Parsons, K. C. 1995b. ISO standards and thermal comfort: Recent development, Chapter 9. In: *Standards for Thermal Comfort: Indoor Air Temperature Standards for the 21st Century*, ed. F. Nicol, M. Humphreys, O. Sykes and S. Roaf. Milton Park, Abingdon, UK: Taylor & Francis, ISBN: O 419 20420 2.

Parsons, K. C. 2000a. Environmental ergonomics: A review of principles, methods and models. *Applied Ergonomics* 31: 581–594.

Parsons, K. C. 2000b. An international thermal comfort standard for people with special requirements. *Proceedings of the Human Factors and Ergonomics Society Annual Meeting* 44(28): 720–723.

Parsons, K. C. 2002. The effects of gender, acclimation state, the opportunity to adjust clothing and physical disability on requirements for thermal comfort. *Energy and Buildings* 34(6): 593–599.

Parsons, K. C. 2003. Thermal comfort for special populations, special environments and adaptive modelling, Chapter 9. In: *Human Thermal Environments: The Effects of Hot, Moderate, and Cold Environments on Human Health, Comfort and Performance*, 2nd edition. London, UK: Taylor & Francis, ISBN: 0-415-23793-9.

Parsons, K. C., B. Shackel and B. Metz. 1995. Ergonomics and international standards: History, organizational structure and method of development. *Applied Ergonomics* 26(4): 249–258.

Psikuta, A., J. Frackiewicz-Kaczmarek, I. Frydrych and R. M. Rossi. 2012. Quantitative evaluation of air gap thickness and contact area between body and garment. *Textile Research Journal* 82(14): 1405–1413.

Rohles, F. H. Jr. and M. A. Johnson. 1972. Thermal comfort in the elderly, NO–2220 (RP-43). *ASHRAE Transactions* 78 (1): 131–137.

Rohles, F. H. Jr. and R. G. Nevins. 1971. The nature of thermal comfort for sedentary man. *ASHRAE Transactions* 77(1): 239–246.

Rohles, F. H., J. E. Woods and R. G. Nevins. 1973. The influence of clothing and temperature on sedentary comfort. *ASHRAE Transactions* 79: 71–80.

Špelić, I., D. Rogale and A. Mihelić-Bogdanić. 2016. An overview of measuring methods and international standards in the field of thermal environment, thermal characteristics of the clothing ensembles and the human subjects' assessment of the thermal comfort. *Tekstil* 65(3–4): 110–122.

Spencer-Smith, J. L. 1977. The physical basis of clothing comfort, part 2: Heat transfer through dry clothing assemblies. *Clothing Research Journal* 5(1): 3–17.

Stolwijk, J. A. J. 1969. *Expansion of a Mathematical Model of thermoregulation to Include High Metabolic Rates*, Final report, NAS-9-7140. Houston, TX: John B. Foundation Laboratory, New Haven for NASA.
Stolwijk, J. A. J. 1970. Mathematical model of thermoregulation, Chapter 48. In: *Physiological and Behavioral Temperature Regulation*, ed. J. D. Hardy, A. P. Gagge and J. A. J. Stolwijlk. Springfield, IL: Charles C Thomas, 703–721.
Stolwijk, J. A. J. 1971. *A Mathematical Model of Physiology Temperature Regulation in Man*. New Haven, CT: Yale University School of Medicine for NASA, NASA CR-1855.
Stolwijk, J. A. J. and J. D. Hardy. 1966a. Temperature regulation in man-a theoretical study. *Pfügers Archives* 291: 129–162.
Stolwijk, J. A. J. and J. D. Hardy. 1966b. Partitional calorimetric studies of response of man to thermal transients. *Journal of Applied Physiology* 21: 967–977.
Studies in Environmental Science 10: Bioengineering, Thermal Physiology And Comfort. 1981. ed. K. Cena and J. A. Clark. Amsterdam, The Netherlands: Elsevier scientific publishing company, ISBN 0-444-99761-X.
Sweeting, C. G. 2015. *United States Army Aviator's Clothing*, 1917–1945. Jefferson, NC: McFarland & Company Inc., e-ISBN: 978-1-4766-1809-8.
Tanabe, S., T. Tsuzuki, K. Kimura and S. Horikawa. 1995. Numerical simulation model of thermal regulation of man with 16 body parts of evaluating thermal environment, part1: Heat transfer at skin surface and comparison with SET* and Stolwijk model. *Summaries of Technical Papers of Annual Meeting Architectural Institute of Japan, D-2: Environmental Engineering* 2: 417–418.
Turner, S. 2008. ASHRAE's thermal comfort standard in America: Future steps away from energy intensive design. In: *Proceedings of Conference: Air Conditioning and the Low Carbon Cooling Challenge*. Windsor, UK: Network for Comfort and Energy Use in Buildings. http://nceub.org.uk.
Vernieuw, C. R., L. A. Stephenson and M. A. Kolka. 2007. Thermal comfort and sensation in men wearing a cooling system controlled by skin temperature. *Human Factors* 49(6): 1033–1044.
Wagner, A. and P. Dorawa. 2016. Research on biophysical properties of protective clothing. *AUTEX Research Journal* 16(4): 236–240.
Wang, F., C. Gao, K. Kuklane, and I. Holmér. 2010. A review of technology of personal heating garments. *International Journal of Occupational Safety and Ergonomics (JOSE)* 16(3): 387–404).
Webb, L. H. and K. C. Parsons. 1998. Case studies of thermal comfort for people with physical disabilities. *ASHRAE Transactions* 104(1): 883–895.
Webb, L. H. and K. C. Parsons. 2000. Thermal comfort requirements for people with physical disabilities. *Human Factors and Ergonomics Society Annual Meeting Proceedings* 44(28):114–121.
Wenzlhuemer, R. 2010. The history of standardisation in Europe. In: *European History Online (EGO)*. Mainz, Germany: the Institute of European History (IEG). http://www.ieg-ego.eu/wenzlhuemerr-2010-en.
WHO Global Disability Action Plan 2014–2021: Better Health for All People with Disability. 2015. Geneva, Switzerland: World Health Organization (WHO), ISBN: 978-92-4-150961-9.
Xu, X. J. and J. Gonzalez. 2011. Determination of the cooling capacity for body ventilation system. *European Journal of Applied Physiology* 111(12): 3155–3160.
Young, A. J., M. N. Sawka and K. B. Pandolf. 1996. Physiology of cold exposure. In: *Nutritional Needs in Cold and in High-Altitude Environments: Applications for Military Personnel in Field Operations, Applications for Military Personnel in Field Operations*. ed. B. M. Marriott and S. J. Carlson. Washington, DC: Nutritional Needs Institute of Medicine (US) Committee on Military Nutrition Research, National Academies Press, ISBN-10: 0-309-05484-2.

Zhang, X. Z., J. Li. and Y. Y. Wang. 2012. Effects of clothing ventilation openings on thermoregulatory responses during exercise. *Indian Journal of Fibre and Textile Research* 37, 162–171.

Zhang, Z.-H., Y. Wang and J. Li. 2011. Model for predicting the effect of an air gap on the heat transfer of a clothed human body. *Fibres & Textiles in Eastern Europe* 19(4–87): 105–110.

Zhang, Z. and J. Li. 2011. Volume of air gaps under clothing and its related thermal effects. *Journal of Fiber Bioengineering & Informatics* 4(2): 137–144.

Zimmermann, C., W. H. Uedelhoven, B. Kurz and K. J. Glitz. 2008. Thermal comfort range of a military cold protection glove: Database by thermophysiological simulation. *European Journal of Applied Physiology* 104: 229–236.

8 The Distribution of Standards on Thermal Comfort

Ergonomics in general is a very complex area and includes a number of areas such as human capabilities and limitations, human–machine interaction, teamwork, tools, machines and material design, environmental factors and work or organisational design. The major fields of ergonomics are physical, psychophysiological, behavioural-cognitive, team, environmental and macroergonomic methods (Stanton, 2005).

The thermal characteristics of an environment are only a small portion of the environmental ergonomics, however, probably the most significant when discussing the ergonomic design of a workplace environment. There is a large variety of objective and subjective methods that help to assess both thermal comfort and strain caused by thermally inadequate environments.

Being able to identify the best practices for analysing the thermal suitability of the environment is not necessarily connected merely to the optimum working conditions. There are some professions that usually work under severe environmental conditions, whether cold or hot. For instance, firefighters that are exposed to heat and fire, or workers in cold storage facilities. Heat production and heat loss are equally important in all conditions, but ratios differ according to the environment. When questioning whether someone is exposed to comfort, heat or cold stress, metabolic heat production and heat loss are always major factors. Cold stress will occur when heat loss exceeds the heat production of human tissues. Contrary to this, heat stress will occur when the body absorbs or produces more heat than can be dissipated through thermoregulatory processes. The balance between metabolic production and heat loss should be close to thermal equilibrium in order to achieve thermal comfort or low thermal strain. When that equilibrium is disturbed illness or even death can occur as the result of the increases or decrease in body core temperature.

The standards in the field of the ergonomics of thermal environments define both the parameters and the measuring methods, which affect human thermoregulation and the prediction of the level of thermal protection and minimum requirements for product quality which serves as protection under extreme temperatures and environmental conditions. Choosing the appropriate measuring method is dependent not only on the property chosen for measurements but also on the environmental conditions and the type of object for which one needs to deduce that specific property.

8.1 THE BASIC DISTRIBUTIONS OF STANDARDS IN THE FIELD OF THE ERGONOMICS OF THE THERMAL ENVIRONMENT

Today, there are over 20 international standards in the field of the ergonomics of the thermal environment, and they are divided into eight categories according to Olesen et al. (Olesen et al., 2016), Figure 8.1:

1. The standards for the hot environments and the assessment of heat stress (ISO 9920, ISO 8996, ISO 7726, ISO 7243, ISO 7933, ISO 9886, ISO 12894, ISO 16595, ISO 13732-1, ISO 15265)
2. The standards for cold environments and the assessment of cold stress ISO 9920, ISO 8996, ISO 7726, ISO 11079, ISO 9886, ISO 12894, ISO 15743, ISO 13732, ISO 15265)
3. The standards for moderate environments (ISO 10551, ISO 9920, ISO 8996, ISO 7726, ISO 7730, ISO 16596, ISO 13732-2, ISO 16594, ISO 14505 1-4, ISO 15265)
4. The standards for vehicles (ISO 14505-2, ISO 14505-3)
5. The standards for people with special requirements (ISO/TS 14415:2005, ISO 28803:2012)

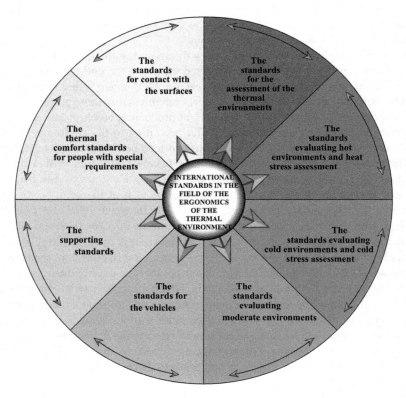

FIGURE 8.1 The basic distributions of the standards in the field of the ergonomics of the thermal environment.

The Distribution of Standards on Thermal Comfort

6. The standards for contact with surfaces (ISO 13732-1, ISO 13732-2, ISO 13732-3)
7. The standards for assessment of the thermal environments (ISO 15265, ISO 15743)
8. Supporting standards (ISO EN 11399, ISO EN 13731, ISO 9920, ISO 8996, ISO 7726, ISO 12894, ISO 9886)

Distribution of the standards can also be based on the type of object for which one needs to deduce some thermally related property, see Figure 8.2:

1. A single layer of fabric, non-woven or the multiple layered textiles (textile composites)
2. The clothing garments or ensembles (requirements for protective clothing)
3. The physiological responses of the human body
4. Physical quantities, environmental indices and strategies

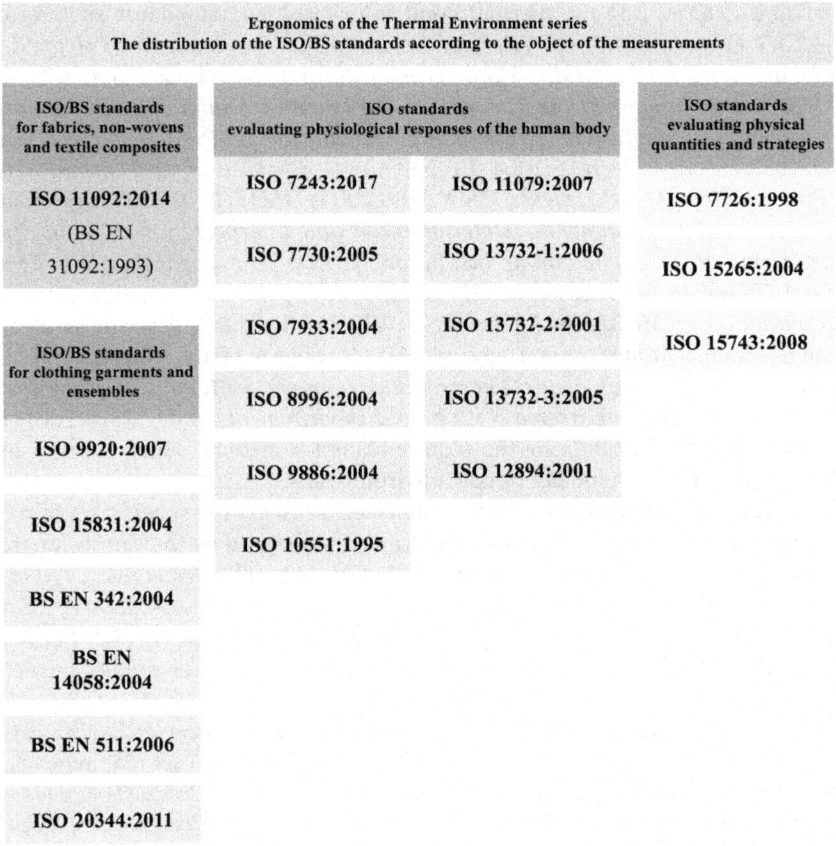

Ergonomics of the Thermal Environment series
The distribution of the ISO/BS standards according to the object of the measurements

ISO/BS standards for fabrics, non-wovens and textile composites	ISO standards evaluating physiological responses of the human body		ISO standards evaluating physical quantities and strategies
ISO 11092:2014 (BS EN 31092:1993)	ISO 7243:2017	ISO 11079:2007	ISO 7726:1998
	ISO 7730:2005	ISO 13732-1:2006	ISO 15265:2004
ISO/BS standards for clothing garments and ensembles	ISO 7933:2004	ISO 13732-2:2001	ISO 15743:2008
	ISO 8996:2004	ISO 13732-3:2005	
ISO 9920:2007	ISO 9886:2004	ISO 12894:2001	
ISO 15831:2004	ISO 10551:1995		
BS EN 342:2004			
BS EN 14058:2004			
BS EN 511:2006			
ISO 20344:2011			

FIGURE 8.2 The distribution of ISO and BS standards according to the object of measurement.

International standards such as ISO 11092:2014 (BS EN 31092:1993) *Physiological Effects – Measurement of Thermal and Water-Vapour Resistance Under Steady-State Conditions (Sweating Guarded-Hotplate Test)* (ISO 11092:2014) is intended for measuring the thermal properties of 2D textile objects, to be exact fabrics, nonwoven cloths and so on. Other standards, such as ISO 9920:2007 *Ergonomics of the Thermal Environment – Estimation of Thermal Insulation and Water-Vapour Resistance of a Clothing Ensemble* (ISO 9920:2007), ISO 15831:2004 *Clothing – Physiological Effects – Measurement of Thermal Insulation by means of a Thermal Manikin* (ISO 15831:2004) and ISO 20344:2011 *Personal Protective Equipment – Test Methods for Footwear* (ISO 20344:2011) aim to determine the thermal properties of 3D objects, meaning single clothing garments or combined garments forming clothing ensembles. One can also choose the standards predicting the impact of the environmental conditions on the human body and its ability to acclimatise. This is ISO 7730:2005 *Ergonomics of the Thermal Environment – Analytical Determination and Interpretation of Thermal Comfort Using Calculation of the PMV and PPD Indices and Local Thermal Comfort Criteria* (ISO 7730:2005), a standard determining the thermal comfort criteria, and standards such as ISO 7243:2017 *Ergonomics of the Thermal Environment – Assessment of Heat Stress Using the WBGT (Wet Bulb Globe Temperature) Index* (ISO 7243:2017), ISO 7933:2004 *Ergonomics of the Thermal Environment – Analytical Determination and Interpretation of Heat Stress Using Calculation of Predicted Heat Strain* (ISO 7933:2004), ISO 9886:2004 *Ergonomics – Evaluation of Thermal Strain by Physiological Measurements* (ISO 9886:2004), ISO 11079:2007 *Ergonomics of the Thermal Environment – Determination and Interpretation of Cold Stress When Using Required Clothing Insulation (IREQ) and Local Cooling Effects* (ISO 11079:2007) and ISO 8996:2004 *Ergonomics of the Thermal Environment – Determination of Metabolic Rate* which evaluates the thermal strain as a result of environmental influence and physiological responses of the human body. ISO 12894:2001 *Ergonomics of the Thermal Environment – Medical Supervision of Individuals Exposed to Extreme Hot or Cold Environments* (ISO 12894:2001) is a supporting standard explaining the requirements for medical supervision of individuals exposed to extreme hot or cold environments.

The requirements set upon clothing intended for usage in a specific set of working conditions and types of activities are primarily assigned by the European group of protective clothing standards such as BS EN 342:2004 *Protective Clothing – Ensembles and Garment for Protection Against Cold* (BS EN 342:2004), BS EN 511:2006 *Protective Clothing – Protective Gloves Against Cold* (BS EN 511:2006) and BS EN 14058:2004 *Protective Clothing – Garment for Protection Against Cool Environments* (BS EN 14058:2004).

The standards such as ISO 7726:1998 *Ergonomics of the Thermal Environment – Instruments for Measuring Physical Quantities* (ISO 7726:1998) set requirements for instruments applicable for measuring physical quantities. ISO 15265:2004 *Ergonomics of the Thermal Environment – Risk Assessment Strategy for the Prevention of Stress or Discomfort in Thermal Working Conditions* (ISO 15265:2004) provides the risk assessment strategy for the prevention of the stress or a discomfort in the thermal working conditions while ISO 15743:2008 *Ergonomics of the Thermal Environment – Cold Workplaces – Risk Assessment and Management* (ISO 15743:2008) provides the risk assessment and management for cold workplaces.

The cold risk management model and its methods form an essential part of ISO CD 15743 *Strategy for Risk Assessment, Management and Work Practice in Cold Environments*. Several ISO standards can be used for assessing thermal risks caused by cold exposure. Most of the methods are self-contained, but they should be used in conjunction with each other in a comprehensive assessment. At present, there are no guidelines on how to systematically assess the effect of cold on human health and performance. The existing guidelines for cold work apply to cold indoor conditions and are limited with regards to practical instructions for the assessment and management of cold. It is important to recognise that, in order for the standards to be utilised as instruments for assessing thermal environments in workplaces, they have to be practically oriented and useable (Risikko et al., 2003).

Another way of the standards' classification is based on the environmental conditions in which the research or an application is performed:

1. The standards for moderate environments
2. The standards for the hot/warm environments
3. The standards for the cold/cool environments

According to Orosa and Oliveira (Orosa and Oliveira, 2012), the main standards applicable for moderate indoor environments are ISO 7726:1998 *Ergonomics of the Thermal Environment – Instruments for Measuring Physical Quantities* (ISO 7726:1998), ISO 7730:2005 *Ergonomics of the Thermal Environment – Analytical Determination and Interpretation of Thermal Comfort Using Calculation of the PMV and PPD Indices and Local Thermal Comfort Criteria* (ISO 7730:2005) and ANSI/ASHRAE 55-2017 *Environmental Conditions for Human Occupancy* (ANSI/ASHRAE Standard 55-2017). Beside these there are other general standards for thermal comfort in moderate environments such as ISO 8996:2004 *Ergonomics of the Thermal Environment – Determination of Metabolic Rate* (ISO 8996:2004), ISO 9920:2008 *Ergonomics of the Thermal Environment – Estimation of Thermal Insulation and Water-Vapour Resistance of a Clothing Ensemble* (ISO 9920:2007), ISO 10551:1995 *Ergonomics of the Thermal Environment – Assessment of the Influence of the Thermal Environment Using Subjective Judgment Scales* (ISO 10551:1995) and ISO 11399:1995 *Ergonomics of the Thermal Environment – Principles and Application of Relevant International Standards* (ISO 11399:1995). The international standards for general thermal comfort in the extreme environments are ISO 7933:2004 *Ergonomics of the Thermal Environment – Analytical Determination and Interpretation of Heat Stress Using Calculation of Predicted Heat Strain* (ISO 7933:2004), ISO 9886:2004 *Ergonomics – Evaluation of Thermal Strain by Physiological Measurements* (ISO 9886:2004) and ISO 11079:2007 *Ergonomics of the Thermal Environment – Determination and Interpretation of Cold Stress When Using Required Clothing Insulation (IREQ) and Local Cooling Effects* (ISO 11079:2007).

The ISO standards in the field of the ergonomics of the thermal environment are divided into three main groups in accordance with the climatic condition (for hot, cold and moderate environments) (Olesen et al., 2016; Parsons, 2008). One can certainly categorise the standards simply in accordance with the object of the measurements. However, since the physiological response of the human body is so complex in nature, it is important to know and to simultaneously use almost all standards for a specific set of environmental conditions. This distribution helps when drafting the testing procedure.

Steps that need to be taken for objective assessment are (Havenith, 2005):

1. Identify and select appropriate equipment for measuring range, accuracy, and response time of apparatus and sensors.
2. Calibrate equipment.
3. Survey expected climate fluctuations in time (seasons, weather) and space within the workspace by interviews with workers.
4. Locate relevant workstations as measurement locations or define locations using a grid of the total workspace.
5. Define locations and measurement timing over the day, season, or weather conditions.
6. Measure and register climatic parameters at three heights (one height in highly uniform environments) at all locations.
7. Survey the workload (metabolic rate) and clothing (insulation) worn at the various workplaces.
8. Correct clothing insulation for movement and wind.
9. Take all data and use heat stress, cold stress, or comfort evaluation methods.

A few of the ISO standards are, however, universal and the foundation for all three categories of the thermal environments. These are: ISO 7726:1998 *Ergonomics of the Thermal Environment – Instruments for Measuring Physical Quantities* (ISO 7726:1998), ISO 8996:2004 *Ergonomics of the Thermal Environment – Determination of Metabolic Rate* (ISO 8996:2004), ISO 9920:2007 *Ergonomics of the Thermal Environment – Estimation of Thermal Insulation and Water-Vapour Resistance of a Clothing Ensemble* (ISO 9920:2008), ISO 9886:2004 *Ergonomics – Evaluation of Thermal Strain by Physiological Measurements* (ISO 9886:2004), ISO 11399:1995 *Ergonomics of the Thermal Environment – Principles and Application of Relevant International Standards* (ISO 11399:1995), ISO 12894:2001 *Ergonomics of the Thermal Environment – Medical Supervision of Individuals Exposed to Extreme Hot or Cold Environments* (ISO 12894:2001) and ISO 15265:2004 *Ergonomics of the Thermal Environment – Risk Assessment Strategy for the Prevention of Stress or Discomfort in Thermal Working Conditions* (ISO 15265:2004). The ISO and European standards for hot environments and the assessment of heat stress are ISO 7243:2017 *Ergonomics of the Thermal Environment – Assessment of Heat Stress Using the WBGT Index* (ISO 7243:2017), also known as BS EN 27243:2017, ISO 7933:2004 *Ergonomics of the Thermal Environment – Analytical Determination and Interpretation of Heat Stress Using Calculation of Predicted Heat Strain*

(ISO 7933:2004) and ISO 13732-1:2006 *Ergonomics of the Thermal Environment – Methods for the Assessment Of Human Responses To Contact With Surfaces – Part 1: Hot Surfaces* (ISO 13732-1:2006) while the measurements in cold environments and the assessment of cold stress are covered by ISO and European standards like ISO 11079:2007 *Ergonomics of the Thermal Environment – Determination and Interpretation of Cold Stress When Using Required Clothing Insulation (IREQ) and Local Cooling Effects* (ISO 11079:2007), ISO 15743:2008 *Ergonomics of the Thermal Environment – Cold Workplaces – Risk Assessment and Management* (ISO 15743:2008), ISO 13732-3:2005 *Ergonomics of the Thermal Environment – Methods for the Assessment of Human Responses to Contact With Surfaces – Part 3: Cold Surfaces* (ISO 13732-3:2005), BS EN 14058:2004 *Protective Clothing – Garment for Protection Against Cool Environments* (BS EN 14058:2004) and BS EN 511:2006 *Protective Clothing – Protective Gloves Against Cold* (BS EN 511:2006). The ISO standards for moderate environments and the assessment of thermal comfort are ISO 7730:2005 *Ergonomics of the Thermal Environment – Analytical Determination and Interpretation of Thermal Comfort Using Calculation of the PMV and PPD Indices and Local Thermal Comfort Criteria* (ISO 7730:2005), ISO 10551:1995 *Ergonomics of the Thermal Environment – Assessment of the influence of the Thermal Environment Using Subjective Judgment Scales* (ISO 10551:1995) and ISO 13732-2:2001 *Ergonomics of the Thermal Environment – Methods for the Assessment of Human Responses To Contact With Surfaces – Part 2: Human Contact with Surfaces at Moderate Temperature* (ISO 13732-2:2001), see Figure 8.3.

8.2 CODING OF STANDARDS FOR CATALOGUING THEIR TYPE

The numbering of the ASTM standards is somewhat different comparing to the numbering and nomenclature of the ISO international standards. While individual ISO standards belong to one separate domain, as for example the domain of *Ergonomics of the Thermal Environments*, ASTM standards belong to the specific section and afterwards to the specific volume. Section 7 covers all the standards intended to be used for the defining and testing of fabrics (around 330 of them) and is divided into two volumes, one of which is 7.01 (fabrics, clothing, textile care, fibres, yarns, non-woven textiles, flammability, accessories such as zippers, etc.) and 7.02 (the body measurements, stitches and seams, UV finishing in textile products, etc.). Section 11 covers the technology of water purification and environmental technology and is composed of eight sub-volumes, with sub-volume 11.03 which is significant to experts in the field of thermal measurements of clothing ensembles. This sub-section covers the topics in the field of Occupational Health and Safety and Protective Clothing. Every ASTM standard can be found in the *Annual Book of ASTM Standards*. The *ASHRAE Handbook of Fundamentals*, on the other hand, contains fundamental data, explanation of the physical quantities and their measuring units (*ASHRAE Handbook of Fundamentals*, 2017). This guide is published in a series containing four volumes and each year one volume is revised to ensure the publishing of all relevant information and actual scientific knowledge. All the corrections can be checked online.

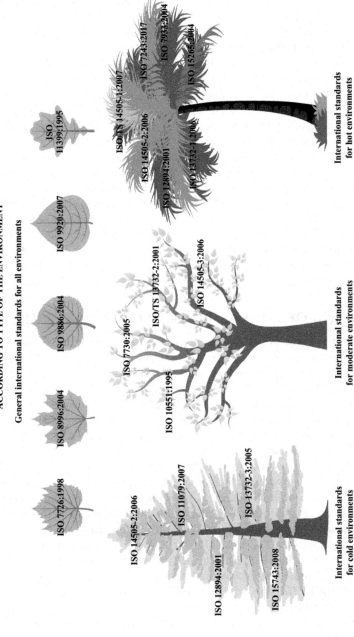

FIGURE 8.3 The distribution of the ISO standards according to environmental conditions.

REFERENCES

ANSI/ASHRAE Standard 55-2017 Thermal Environmental Conditions for Human Occupancy. 2017. Atlanta, USA: American Society of Heating, Refrigerating and Air Conditioning Engineers Inc. (ASHRAE), ISSN: 1041-2336.

ASHRAE Handbook Fundamentals (SI ed.). 2017. Atlanta, USA: American Society of Heating, Refrigerating, and Air Conditioning Engineers Inc., ISBN: 1931862702.

BS EN 14058:2004 Protective Clothing – Garment for Protection against Cool Environments. 2004. BSI.

BS EN 342:2004 Protective Clothing – Ensembles and Garment for Protection against Cold. 2004. BSI.

BS EN 511:2006 Protective Clothing – Protective Gloves against Cold. 2006. BSI.

Havenith, G. 2005. Thermal conditions measurement, Chapter 60. In: *Handbook of Human Factors and Ergonomics Methods*, eds. N. Stanton, A. Hedge, K. Brookhuis, E. Salas and H. Hendrics. Boca Raton, FL: CRC Press, 552–573, ISBN: 0-415-28700-6.

ISO 10551:1995 Ergonomics of the Thermal Environment – Assessment of the Influence of the Thermal Environment using Subjective Judgment Scales. 1995. ISO-International Organization for Standardization. (also BS EN ISO 10551:2001).

ISO 11079:2007 Ergonomics of the Thermal Environment – Determination and Interpretation of Cold Stress When using Required Clothing Insulation (IREQ) and Local Cooling Effects. 2007. ISO-International Organization for Standardization.

ISO 11092:2014 Physiological Effects – Measurement of Thermal and Water-Vapour Resistance under Steady-State Conditions (Sweating Guarded-Hotplate Test). 2014. ISO-International Organization for Standardization.

ISO 11399:1995 Ergonomics of the Thermal Environment – Principles and Application of Relevant International Standards. 1995. ISO-International Organization for Standardization.

ISO 12894:2001 Ergonomics of the Thermal Environment – Medical Supervision of Individuals Exposed to Extreme Hot or Cold Environments. 2001. ISO-International Organization for Standardization.

ISO 13732-1:2006 Ergonomics of the Thermal Environment – Methods for the Assessment of Human Responses to Contact with Surfaces – Part 1: Hot Surfaces. 2004. ISO-International Organization for Standardization.

ISO 13732-2:2001 Ergonomics of the Thermal Environment – Methods for the Assessment of Human Responses to Contact with Surfaces – Part 2: Human Contact with Surfaces at Moderate Temperature. 2001. ISO-International Organization for Standardization.

ISO 13732-3:2005 Ergonomics of the Thermal Environment – Methods for the Assessment of Human Responses to Contact with Surfaces – Part 3: Cold Surfaces. 2005. ISO-International Organization for Standardization.

ISO 15265:2004 Ergonomics of the Thermal Environment – Risk Assessment Strategy for the Prevention of Stress or Discomfort in Thermal Working Conditions. 2004. ISO-International Organization for Standardization.

ISO 15743:2008 Ergonomics of the Thermal Environment – Cold Workplaces – Risk Assessment and Management. 2008. ISO-International Organization for Standardization.

ISO 15831:2004 Clothing – Physiological Effects – Measurement of Thermal Insulation by Means of a Thermal Manikin. 2004. ISO-International Organization for Standardization.

ISO 20344:2011 Personal Protective Equipment – Test Methods for Footwear. 2011. ISO-International Organization for Standardization.

ISO 7243:2017 Ergonomics of the Thermal Environment – Assessment of Heat Stress using the WBGT (Wet Bulb Globe Temperature) Index. 2017. ISO-International Organization for Standardization. (BS EN 27243:2017).

ISO 7726:1998 Ergonomics of the Thermal Environment – Instruments for Measuring Physical Quantities. 1998. ISO-International Organization for Standardization.

ISO 7730:2005 Ergonomics of the Thermal Environment – Analytical Determination and Interpretation of Thermal Comfort using Calculation of the PMV and PPD Indices and Local Thermal Comfort Criteria. 2005. ISO-International Organization for Standardization.

ISO 7933:2004 Ergonomics of the Thermal Environment – Analytical Determination and Interpretation of Heat Stress using Calculation of Predicted Heat Strain. 2004. ISO-International Organization for Standardization.

ISO 8996:2004 Ergonomics of the Thermal Environment – Determination of Metabolic Rate. 2004. ISO-International Organization for Standardization.

ISO 9886:2004 Ergonomics – Evaluation of Thermal Strain by Physiological Measurements. 2004. ISO-International Organization for Standardization.

ISO 9920:2007 Ergonomics of the Thermal Environment – Estimation of Thermal Insulation and Water-Vapour Resistance of a Clothing Ensemble. 2007. ISO-International Organization for Standardization.

Olesen, B. W., F. R. d'Ambrosio Alfano, K. Parsons and B. I. Palella. 2016. The history of international standardization for the ergonomics of the thermal environment. In: *The Book of Proceedings of the 9th Windsor Conference: Making Comfort Relevant*, eds. L. Brotas, S. Roaf, F. Nicoland, and M. Humphreys. Windsor, UK: Network for Comfort and Energy Use in Buildings (NCEUB), 15–38, ISBN: 978-0-9928957-3-0.

Orosa, J. A. and A. C. Oliveira. 2012. *Passive Methods as a Solution for Improving Indoor Environments. Green Energy and Technology*. London, UK: Springer-Verlag, ISBN: 978-1-4471-2336-1.

Parsons, K. C. 2008. Industrial health for all: Appropriate physical environments, inclusive design, and standards that are truly international. *Industrial Health* 46: 195–197.

Risikko, T., T. M. Mäkinen, A. Påsche, L. Toivonen and J. Hassi. 2003. A model for managing cold-related health and safety risks at workplaces. *International Journal of Circumpolar Health* 62(2): 204–215.

Stanton, N. 2005. Human factors and ergonomics methods, Chapter 1. In: *Handbook of Human Factors and Ergonomics Methods*, eds. N. Stanton, A. Hedge, K. Brookhuis, E. Salas and H. Hendrics. Boca Raton, FL, USA: CRC Press, 552–573, ISBN: 0-415-28700-6.

9 Overview of the Most Significant Standards on Thermal Comfort

9.1 THE MOST SIGNIFICANT ISO AND EUROPEAN STANDARDS ON THERMAL COMFORT

The international standards covered by the area of the ergonomics of the thermal environment can be categorised according to the field of use, or respectively the thermal properties of the environment described. The system is categorised into standards relating to cold stress, heat stress and thermal comfort, which will be explained in detail in the following.

The first and prime working group to deal with the creation of standards on thermal comfort is the working group WG1 *Ergonomics of the Thermal Environment*, which falls under the structure of the technical committee ISO/TC 159 *Ergonomics* and the subcommittee SC 5 *Ergonomics of the Physical Environment*. However, standards on the test methods for obtaining the thermal properties of textiles and clothing, which are used in the research on thermal comfort, fall under the scope of technical committee ISO/TC 38 *Textiles* (ISO website, ISO/TC 38) and ISO/TC 94 *Personal Safety – Personal Protective Equipment* (ISO website, ISO/TC 94), which is further divided to the subcommittee SC 13 *Protective Clothing* (ISO website, ISO/TC 94/SC 13) and SC 3 *Foot Protection* (ISO website, ISO/TC 94/SC 3), Figure 9.1.

The important ISO/BS standards in the field of ergonomics of the thermal environment are shown in Table 9.1.

The supporting standards cover all of the other significant factors, which can be related in order for humans to feel thermally comfortable within their surroundings. These factors are the metabolic rate and thermal strain (ISO 8996:2004; ISO 9886:2004), clothing (ISO 9920:2007; ISO 15831:2004), the overall risk assessment strategy relevant to physical environments (ISO 15265:2004) and the subjective ratings, whether obtained by laboratory trials (ISO 10551:1995) or environmental surveys (ISO 28802:2012).

As previously mentioned, there are no ISO standards on clothing for protection against cool and cold environments, but there are plenty of standards on clothing for protection against heat and flame. However, there are standards on clothing for protection against cool and cold from the British Standards Institution (BSI), which can be used in the absence of the official international standard. There are also no international standards which can define the requirements on everyday clothing that serves as protection against more severe environmental conditions. This leaves buyers to rely on their own expertise and experience. Growing consumer tastes and a

Standard Methods for Thermal Comfort Assessment of Clothing

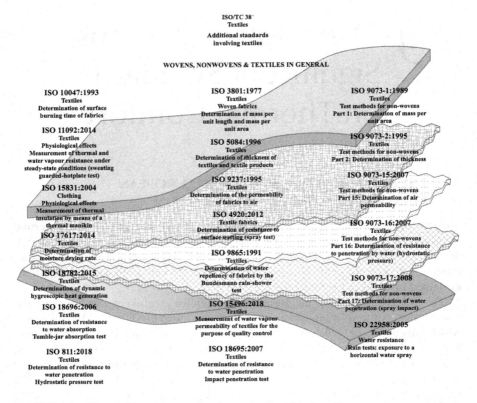

FIGURE 9.1 The additional international standards which describe measurement methods for thermal, moisture and other related properties of textiles.

competitive market will certainly lead to the creation of standards covering the thermal and associated properties of textile/clothing for the general public.

9.1.1 The Standard Evaluating Physical Quantities

External parameters that need to be assessed to determine heat or cold stress levels or comfort conditions include temperature, air humidity and airspeed (Havenith, 2005).

Directly measured external parameters are (ASHRAE Handbook Fundamentals, 2017):

1. Air temperature (t_a)
2. Wet bulb temperature (t_{wb})
3. Dew-point temperature (t_{dp})
4. Water vapour pressure (p_a)
5. Total atmospheric pressure (p_t)
6. Relative humidity (r_h)
7. Humidity ratio (W_a)

TABLE 9.1
The Overview of ISO/BS Standards in the Field of Thermal Measurements, Defining the Thermal Environment and Thermal Request Set upon the Textiles and the Clothing

Designation	Title
ISO 7726:1998	*Ergonomics of the Thermal Environment – Instruments for Measuring Physical Quantities*
ISO 7730:2005	*Ergonomics of the Thermal Environment – Analytical Determination and Interpretation of Thermal Comfort Using a Calculation of the PMV and PPD Indices and Local Thermal Comfort Criteria*
ISO 7933:2004	*Ergonomics of the Thermal Environment – Analytical Determination and Interpretation of Heat Stress Using a Calculation of Predicted Heat Strain*
ISO 8996:2004	*Ergonomics of the Thermal Environment – Determination of Metabolic Rate*
ISO 9886:2004	*Ergonomics – Evaluation of Thermal Strain by Physiological Measurements*
ISO 9920:2007	*Ergonomics of the Thermal Environment – Estimation of Thermal Insulation and Water-Vapour Resistance of a Clothing Ensemble*
ISO 10551:1995	*Ergonomics of the Thermal Environment – Assessment of the Influence of the Thermal Environment Using Subjective Judgement Scales*
ISO 11079:2007	*Ergonomics of the Thermal Environment – Determination and Interpretation of Cold Stress When Using Required Clothing Insulation (IREQ) and Local Cooling Effects*
ISO 11092:2014 (BS EN 31092:1993)	*Textile – Physiological Effects – Measurement of Thermal and Water-Vapour Resistance Under Steady-State Conditions (Sweating Guarded-Hotplate Test)*
BS EN 342:2004	*Protective Clothing Ensembles and Garment for Protection Against Cold*
BS EN 511:2006	*Protective Gloves Against Cold*
BS EN 14058:2004	*Protective Clothing Garment for Protection Against Cool Environments*
ISO 15831:2004	*Clothing – Physiological Effects – Measurement of Thermal Insulation by Means of a Thermal Manikin*
ISO 20344:2011	*Personal Protective Equipment – Test Methods for Footwear*

Calculated parameters:

1. Mean radiant temperature (t_r)
2. Plane radiant temperature (t_{pr})

ISO 7726:1998 *Ergonomics of the Thermal Environment – Instruments for Measuring Physical Quantities* specifies the minimum characteristics of instruments and measuring methods for measuring physical quantities characterising different environments. This norm doesn't define an overall index of the thermal comfort or the thermal stress, but it standardises the process of recording information leading to the determination of such indices subsequently. Since the increase in heat loss is induced by surrounding temperature enlargement, it is crucial to monitor the air temperature, the mean radiant temperature and the surface temperatures, if required. The basic physical quantities derived by this specific standard are air temperature, mean radiant temperature, the absolute humidity of the air, air velocity and surface temperature (ISO 7726:1998).

The measurement methods defined by this standard are classified into two categories according to the predicted thermal environment. The C category specifies the conditions and the methods that relate to measurements carried out in moderate environments approaching comfort conditions, while the S category specifies the conditions and the methods that relate to measurements carried out in environments subjected to greater thermal stress or even extreme thermal stress.

The physical quantities and corresponding measuring methods explained by this standard are:

1. The air temperature (t_a)
2. The mean radiant temperature (t_r)
3. The radiant plane temperature (t_{pr})
4. Partial pressure of the water vapour also called the absolute humidity of the air (p_a)
5. The air velocity (v_a)
6. The operative temperature (t_o)
7. The surface temperature (t_s)

An environment may be considered as homogeneous from a bio-climatic point of view if, at a given moment, the air temperature, the radiation, the air velocity and the humidity can be considered to be practically uniform around the subject. When the environment is too heterogeneous, the physical quantities should be measured at several locations around the subject, and the mean values should be assessed. The derived physical quantities characterise a group of environmental factors, weighted according to the characteristics of the sensors used. They are often used to define an empirical index of comfort or thermal stress without having recourse to a rational method based on estimates of the various forms of heat exchange between the human body and the thermal environments, and of the resulting thermal balance and the physiological strain.

9.1.1.1 Temperature Measurements – Requirements for Measuring Equipment

Practitioners should be aware of equipment demands when measuring certain quantities. For example, when measuring the air temperature of the space (also called the ambient temperature, t_a), one has to bear in mind the minimum sensor requirements. The sensor range for measuring comfort must be 283.15–313.15/±0.5 K (10–40°C/±0.5°C). The sensor range for measuring stress is 233.15–393.15/±1 K (−40–120°C/±1°C) with a desirable accuracy of ±0.2 (Havenith, 2005).

The temperature can be measured in accordance with the application and the approximate range by various temperature sensors. It can be measured by contact and non-contact temperature sensing detectors. Any device that changes monotonically with temperature is a thermometer, but the term usually stands for an ordinary liquid-in-glass thermometer. Each temperature sensor should be calibrated and its recovery factor determined in order to obtain the precise temperature measurement. For most temperature sensors applicable to air velocities below 10 m/s, the recovery factor can be ignored (*ASHRAE Handbook of Fundamentals*, 2017).

The measurements done for mean radiant temperatures also require appropriate equipment. The sensor range for measuring inside the comfort zones should follow the range of 283.15–313.15/±2 K (10–40/±2°C) with a desirable accuracy for comfort of ±0.2. When questioning the thermal stress, the sensor range should be from 233.15–323.15/±5 K (−40–50/±5°C) with a desirable accuracy for stress of ±0.5. Mean radiant temperatures measured above 323.15 K (50°C) requires a sensor range with a linear increase from 278.15 to 286.15 K (5 to 13°C) at 423 K (150°C). For measurement of plane radiant temperatures, typically used in comfort assessment, the range falls between 273.15 to 323.15/±0.5 K (0 to 50/±0.5°C) (Havenith, 2005).

The surface temperatures can be measured by two types of sensors. One of them is the special sensors that ensure good contact with the surface while insulating the sensor from the environment. The conduction between surface and sensor must be much higher than that from the sensor to the environment. The other is a non-contact infrared sensor. The surfaces with very low conductivity may yield false values, and in such cases a non-contact infrared sensor is better. Accuracy is dependent on the emissivity/reflectivity of the surface. The sensor for measuring inside the comfort zones requires a range of 273.15 to 323.15/±1 K (0 to 50/±1°C) with a desirable accuracy comfort of ±0.5. The sensor required for measuring stress is obliged to have a range from 233.15 to 393.15 K (−40–120°C) with a desirable accuracy for stress of ±0.5. There are two basic sub-rangers, one being between 263.15 and 323.15/±1 K (−10 and +50/±1°C), and the other below 263.15 K and above 323.15 K (−10°C and above +50°C), which requires a linear increase from 1 to 3.5°C and 4.5°C to the range limit (Havenith, 2005).

These are (*ASHRAE Handbook of Fundamentals*, 2017; Havenith, 2005):

1. Thermometers:
   - Liquid-in-glass thermometers (temperature of gases and liquids by contact): These are used for applications such as the local temperature indication of process fluids for total or partial stem immersion with relatively high accuracy at low cost.
   - Gas thermometers (remote readings of gases temperature by contact): These thermometers can be affected by radiation from surrounding surfaces. The radiation effects can be ignored if the gas temperature is approximately the same as those of the surrounding surfaces. If there are temperature differences, radiation effects should be minimised by shielding (placing the highly reflective surfaces between the thermometer bulb and the surrounding surfaces such that air movement around the bulb is not restricted) or aspiration (passing a high-velocity stream of air or gas over the thermometer bulb).

- Resistance thermometers (remote readings by contact): Their measuring process depends on a change of the electrical resistance of a metal sensing element with a change in temperature and according to their ability to give better results when used to measure steady or slowly changing temperature. They have longer response times than thermocouples and are more costly to make. There are two major types: resistance temperature devices (simple circuit designs, high degree of linearity, good sensitivity and excellent stability) and thermistors (semiconductor, with the ability to exhibit large changes in resistance with temperature, connected by lead wires into a galvanometer bridge circuit; usually applied to electronic temperature compensation circuits, where high resolution and limited operating temperature ranges exist).
- Metastatic thermometers (remote readings by contact).
- Bimetallic thermometers (due to time lag unsuitable for remote use).
- Pressure-bulb thermometers (remote testing).

2. Thermocouples (remote readings by contact and used for surface temperature measurement): Constructed by two wires of dissimilar metals joined together (thermojunction; working principle is the accurate measurement of the Seebeck voltage). They are less precise than platinum resistance thermometers; however, due to low cost, moderate reliability, and ease of use, they are the most common instruments of temperature measurement for the range of 273.15 to 1273.15 K (0 to 1000°C).

3. Pyrometers or infrared radiation thermometers (remote-sensing non-contact thermometer, a remote temperature sensor measures the surface temperature due to sensing the thermal radiation emitted from the surface over a wide range of temperatures): The radiant flux from the observed object is focused by an optical system onto an infrared detector that generates an output signal proportional to the incident radiation that can be read from a meter or display unit. A radiometer only measures the power level of the radiation incident on the detector; this incident radiation is a combination of the thermal radiation emitted by the object and the surrounding background radiation reflected from the surface of the object. There are point and scanning radiometers. These radiometers detectors are classified according to the detector as thermal (thermopile and metallic and semiconductor bolometers) or photon. The thermal detectors detect if a change in the electrical property is caused by the heating effect of the incident radiation. The photon detectors (scanning radiometers usually use photon detectors) detect a change in electrical property caused by the surface absorption of incident photons (shorter response time). The mean radiant temperature is the uniform temperature of an imaginary black enclosure in which an occupant would exchange the same amount of radiant heat as in the actual non-uniform enclosure. It can be calculated from measured surface temperatures and the corresponding angle factors between the person and the surfaces or from the plane radiant temperature in six opposite directions weighted according to the projected area factors for a person (*ASHRAE Handbook of Fundamentals*, 2017; Havenith, 2005).

9.1.1.2 HUMIDITY MEASUREMENTS – REQUIREMENTS FOR MEASURING EQUIPMENT

Equipment requirements should also be checked when measuring other quantities. The relative humidity can be measured with hair hygrometers, but these have very limited accuracy and react slowly. On the other hand, the electronic sensors are usually cheaper, but when exposed to extreme climates they need to be regularly recalibrated. The more expensive solution is found when using dew-point sensors, which are very accurate. There are also psychrometers, which are both accurate and affordable (Havenith, 2005).

Humidity is measured by a hygrometer, which is an instrument capable of measuring the humidity or psychrometric state of air where sensors respond to different moisture contents (wet bulb temperature, relative humidity, humidity mixing ratio, dew-point and frost point). There are different hygrometers for certain conditions and within specific limitations. Almost every psychrometer consists of a pair of matched electrical or mechanical temperature sensors, one of which is moistened, and the blower aspirates the sensor, lowering the temperature at the moistened temperature sensor. This is a steady-state, open-loop, non-equilibrium process, which depends on the purity of the water, the cleanliness of the wick, the ventilation rate, radiation effects, the size and accuracy of the temperature sensors, and the transport properties of the gas (*ASHRAE Handbook of Fundamentals*, 2017; Havenith, 2005).

Some of the hygrometers' categories due to the working principle are (*ASHRAE Handbook of Fundamentals*, 2017; Havenith, 2005):

1. Hygrometers working on the basis of evaporative cooling that measures the temperature of wet bulbs such as a classical psychrometer or adiabatic saturation psychrometer.
2. The condensation (chilled mirror) dew-point hygrometer works on the basis of optically determining the moisture formation. It is accurate and reliable within a wide humidity range; however, it is very complex and costly. The surface is cooled (thermoelectrically, mechanically, or chemically) until dew or frost begins to condense out. The condensate surface is maintained electronically in vapour pressure equilibrium with the surrounding gas, while surface condensation is detected by optical, electrical or nuclear techniques. The measured surface temperature is then the dew-point temperature. The dew-point temperature is a fundamental measure of air humidity defined as the temperature at which condensation first occurs when an air–water vapour mixture is cooled at constant pressure.
3. Hygrometers work on the basis of the water vapour pressure depression in salt solution where the sensor is in the heated saturated salt solution (self-heating salt-phase transition hygrometer or a heated electrical hygrometer). It consists of a tubular substrate covered by glass fibre fabric, with a spiral bifilar winding for electrodes. The surface is covered with a salt solution, and the sensor is connected in series with a 24 V supply. This electrical current flowing through the salt film heats the sensor. When the balance is reached, with the water vapour pressures of the salt film and ambient atmosphere equal, the sensor temperature adjusts automatically.

4. Hygrometers with mechanical sensors such as hair, nylon, Dacron thread, Goldbeater's skin, cellulosic materials or carbon, which measure the dimensional change of the thread due to moisture.
5. Hygrometers with electrical sensors such as ion exchange resin, porous ceramic, aluminium oxide and electrolytic hygrometer, which measure the impedance or capacitance change due to humidity. There are also the infrared laser diode (optical diodes) and surface acoustic wave (surface acoustic wave attenuation) sensors.
6. Coulometric electrolytic hygrometers, which measure mass changes due to adsorbed moisture. The air passes through a tube where moisture is adsorbed by a phosphorous pentoxide and electrolysed into hydrogen and oxygen. The current of electrolysis determines the mass of water vapour entering the sensor.
7. Hygrometers with mass sensitive piezoelectric sensors, which measure mass changes due to adsorbed moisture. This hygrometer compares the changes in the frequency of two hygroscopically coated quartz crystal oscillators. The amount of water absorbed into the coated quartz crystal oscillator is a function of relative humidity. The water vapour absorption changes the mass of the crystal and correspondingly the frequency.
8. Spectroscopic (radiation absorption) hygrometers that measure moisture absorption of UV or IR radiation. The hygrometer consists of an energy source, an optical system for isolating wavelengths due to selective absorption of radiation (water vapour absorbs infrared radiation at 2 to 3 μm and ultraviolet radiation at 0.122 μm wavelength) and a measurement system for determining the attenuation of radiant energy caused by the water vapour in the optical path. The absorbed radiation is measured extremely quickly and independent of the degree of saturation of the gas mixture.
9. Gravimetric hygrometers that directly measure the mixing ratio by comparing the sample gas with dry airstream due to extracting the mass of water vapour in a known quantity or atmosphere. The gravimetric hygrometer gives the absolute water vapour content, where the mass of the absorbed water and the precise measurement of the gas volume associated with the water vapour determine the mixing ratio or absolute humidity of the sample with extreme precision. However, this method is complex and requires much attention to detail.
10. Hygrometers that measure colour changes.

9.1.1.3 Pressure Measurements – Requirements for Measuring Equipment

The absolute pressure is the force exerted per unit area by a medium (usually required for thermodynamic and material properties). Vacuum refers to pressures below atmospheric pressure value. Differential pressure is the difference between two absolute pressures (gauge pressure). Pressures are also classified as static or dynamic. Static pressures have a small or undetectable change with time, while the dynamic pressures include a significant pulsed, oscillatory, or other time-dependent components (*ASHRAE Handbook of Fundamentals*, 2017).

The pressure measuring instruments are divided into three groups: standards, mechanical gauges and electromechanical transducers. Standards instruments are the most accurate, mechanical are the cheapest and the most common, while electromechanical transducers have become much less expensive and are easier to use. Accurate values of atmospheric or barometric pressure are required for weather prediction and aircraft altimetry (*ASHRAE Handbook of Fundamentals*, 2017).

The types of pressure measuring instruments (*ASHRAE Handbook of Fundamentals*, 2017):

1. Standards
   - Liquid-column manometers measure pressure by determining the vertical displacement of a liquid of known density in a known gravitational field (U-tube). The pressure to be measured is applied to one side of the U-tube. If the other reference side is evacuated above zero pressure, the manometer measures absolute pressure; if the reference side is open to the atmosphere, it measures gauge pressure; if the reference side is connected to some other pressure, the manometer measures the differential between the two pressures.
   - Piston-gauge, pressure-balance, or deadweight-tester type are standards for pressures above the range of manometers. These instruments apply pressure to the bottom of a vertical piston (surrounded by a close-fitting cylinder), which generates a force approximately equal to the pressure times the area of the piston. The force is balanced by weights stacked on the top of the piston. If the mass of the weights, the local acceleration of gravity and the effective area of the piston and cylinder assembly are known, the applied pressure can be calculated.
   - The low-pressure standard such as the McLeod gauge measure very low absolute pressures (below about 100 Pa).
2. Mechanical pressure gauges: Mechanical pressure gauges couple a pressure sensor to a mechanical readout (pointer and dial). The Bourdon tube sensor, a coiled metal tube of a circular or elliptical cross-section, works by applying pressure to the inside of the tube and causing it to uncoil. A mechanical linkage translates the motion of the end of the tube to the rotation of a pointer. They are limited by inelastic behaviour of the sensing element, friction in the readout mechanism and limited resolution of the pointer and dial.
3. Electromechanical transducers: Electromechanical transducers reduce the limitations of the mechanical pressure gauges by using electronic techniques to sense the distortion or stress of a mechanical sensing element and electronically convert that stress or distortion to a pressure reading. They are usually used to measure pressure below about 10 kPa. The feedback techniques may be used to constrain the sensor in a null position, minimising distortion and hysteresis of the sensing element, as well as the temperature control or compensation.

9.1.1.4 Air Velocity Measurements – Requirements for Measuring Equipment

Measurements of air velocity are usually made by timing the rate of movement of solid tracers or by monitoring the change in concentration level of gas tracers (*ASHRAE Handbook of Fundamentals, 2017*; Havenith, 2005).

Whether the measurements of spatial or temporal variations in airspeed are conducted, few anemometer types can be used. The vane or cup anemometer is used if the air movement is coming from a fixed direction and fluctuates only slowly. A hot-wire anemometer is used if the wind is not unidirectional or fluctuates rapidly. The air velocity can also be measured with a Kata-thermometer. The sensor requirements are set for comfort in the range of 0.05 to $1 \pm (0.05 + 0.05)$ m/s with accuracy (expressed as the response time, where 90% of final value is reached) being ≤ 0.5 s. The sensor requirements are set for stress in the range of 0.2 to $20 \pm (0.1 + 0.05)$ ms^{-1} with the response time ≤ 0.2 s (for measurement of turbulence intensity) (Havenith, 2005).

Anemometers are divided into deflecting vane anemometers, propeller or revolving vane anemometers, cup anemometers and thermal or hot-wire anemometers. Low and highly directional air velocities in rooms are measured by a smoke puff or solid airborne tracer. Directional moderate air velocities in rooms can also be measured by the deflecting vane anemometer or revolving vane anemometer, while the low and high air velocities can be measured by the hot-wire anemometer. Pitot tubes are used to measure duct standard and transient velocity with turbulence. The pitot-static tube, in conjunction with a suitable manometer or differential pressure transducer, provides a simple method of determining air velocity at a point in a flow field (usually velocities above 7.5 m/s). Other high-velocity measurements include the impact tube. Meteorological measurements are done by the cup anemometer, which has poor accuracy at low air velocity below 2.5 m/s.

There are also the laser Doppler velocimeters or anemometers, which are extremely complex and collect scattered light produced by a particle passing through the intersection volume of two intersecting laser beams of the same light frequency (*ASHRAE Handbook of Fundamentals*, 2017).

9.1.2 Standards Assessing the Thermal Comfort and Physiological Responses of Humans

The standards ISO 7726 (ISO 7726:1998) (measuring physical quantities) and ISO 8996 (ISO 8996:2004) (determination of metabolic rate) are used to describe and quantify the parameters influencing human thermoregulation in some specific environment, but standards such as ISO 9886 (ISO 9886:2004) (evaluation of thermal strain by physiological measurements) and ISO 9920 (ISO 9920:2007) (estimation of thermal insulation and water vapour resistance of a clothing ensemble) describe how to predict the degree of thermal protection based on those parameters, apropos the thermal comfort or discomfort and health risks possibly sustained by the subjects in certain environment.

The methods predicting general thermal sensation and degree of discomfort (thermal dissatisfaction) of people exposed to the moderate thermal environments are described in standard ISO 7730:2005 *Ergonomics of the Thermal Environment – Analytical Determination And Interpretation of Thermal Comfort Using a Calculation of the PMV and PPD Indices and Local Thermal Comfort Criteria*. This method enables the analytical determination and interpretation of thermal comfort using a calculation of PMV (predicted mean vote) and PPD (predicted percentage of dissatisfied) and local thermal comfort criteria (ISO 7730:2005).

It summarises the environmental conditions that are considered acceptable for the general thermal comfort as well as those representing local discomfort, caused by the thermal environment. It is applicable to healthy men and women exposed to (un)comfortable indoor environments, depending whether one wants to conduct an assessment for improving the existing conditions or design new ones. It is intended to be used with reference to ISO/TS 14415:2005 (ISO 14415:2005) when considering persons with the special requirements. This standard is currently withdrawn and replaced by ISO 28803:2012 *Ergonomics of the Physical Environment, Application of International Standards to People with Special Requirements* (ISO 28803:2012). Comfort limits can be expressed by the PMV and PPD indices by calculating local discomfort factors such as draught, vertical air temperature difference, cold or warm floors and radiant temperature asymmetry. It describes the environmental conditions considered acceptable for general thermal comfort as well as for the non-steady-state thermal environments. The concept of radiant temperature asymmetry is used when the radiative environment cannot be not completely described by the mean radiant temperature (*ASHRAE Handbook of Fundamentals*, 2017).

For PMV model data on the climate, clothing and metabolic heat are required. Havenith et al. (Havenith et al., 2002) proposed a revision of the former PMV model from 1995, incorporating corrections of clothing heat resistance, clothing water vapour resistance and metabolic heat production. The two studies (Havenith et al., 1999; Holmér et al., 1999) proposed the correction of clothing insulation and vapour resistance due to air movement and wind motion. Second, Havenith et al. (Havenith et al., 2002) concluded that improvements in the metabolic rate estimation in ISO 8996 are needed. Using the values for the metabolic rate in ISO 8996 can lead to an overestimation of the cold stress, since they do not consider the increase in metabolic rate by the protective clothing (Kuklane et al., 2007).

The standard ISO 8996:2004 *Ergonomics of the Thermal Environment – Determination Of Metabolic Rate*) (ISO 8996:2004) is used to determine the metabolic rate, as a conversion of the chemical into the mechanical and the thermal energy. This is an important supporting standard for metabolic heat estimation and is used in compliance with ISO 7243:2017 (ISO 7243:2017) and ISO 7933:2004 (ISO 7933:2004).

Determining metabolic rates is important because almost all metabolic energy is released as heat in the body. ISO 8996:2004 presents both methods for estimating metabolic rate. It can be measured using indirect calorimetry (measuring oxygen uptake) or using tables describing activities, professions and postures (ISO 8996:2004).

But one should always be cautious because the estimations of metabolic rate and of clothing insulation can show large errors (±10%). This should be considered when evaluating the outcome of the analyses (Havenith, 2005). This is the reason why thermal manikin measurements became the commonly used practice when determining clothing insulation which will be later presented when discussing ISO 15831:2004 (ISO 15831:2004).

It describes the measurements of the energetic cost of a muscular load and gives a numerical index of activity. Metabolic rate is the crucial factor while determining comfort or strain resulting from exposure to the thermal environment. In hot environments, the thermal strain rises due to the high levels of metabolic heat production associated with muscular work. The large amounts of heat need to be dissipated from the body into the environment, mostly by sweating. The data included in this standard concerns an average male, 1.75 m high, weighing 70 kg with a body surface area of 1.8 m^2 and average female, 1.70 m high, weighing 60 kg with a body surface area of 1.6 m^2. This standard implies that the total energy consumption while working is assumed equal to heat production. The mechanical efficiency of muscular work, so-called useful work, W, is negligible.

Four approaches for determining the metabolic rate are given by the standard. The choice is given between two methods: Method 1A is a classification according to occupation while method 1B is a classification according to the kind of activity. Method 2A is the method of determining the metabolic rate by adding the metabolic rate for body posture or by adding the metabolic rate, as the body is in motion and relating to work speed, to the baseline metabolic rate. Method 2B is the method of determining the metabolic rate by means of tabulated values for various activities. Metabolic rate could be determined from heart rate recordings over a representative period of time (level 3 analysis). This method for the indirect determination of the metabolic rate is based on the relationship between the oxygen uptake and the heart rate under defined conditions. The trained experts perform the expertise in order to assess the metabolic rate by choosing three possible methods of the assessment. Method 4A covers the oxygen consumption measured over short periods of time (10 to 20 minutes) as opposed to the 4B method, the so-called doubly labelled water method, which aims to characterise the average metabolic rate over much longer time periods (one to two weeks). Method 4C is a direct calorimetry method.

According to Parsons certain improvements of this standard are needed, including the adjustment of metabolic rates for the types of the indoor or the outdoor work, customisation for different cultures and populations, customisation for the difference between the paced work, maintaining the same level of activity, and non-paced work with the activity level variation, customisation for the influence of the clothing and the impact of the individual differences on the metabolic rate for the same activity (Parsons, 2013).

The thermal manikins are used in the clothing industry for determining clothing and sleeping bag thermal insulation, as well as the parameters of the surrounding environment in which the use of a sleeping bag will not cause cooling or overheating of the body. They are also used to assess thermal comfort during work performed in clothes in a given thermal environment, to test heat and water vapour transfer through clothing, for testing clothing insulation and clothing evaporative resistance, to assess the cooling capacity of phase transformation compounds used in clothes which are designed for performing work in a hot climate and to model thermal insulation on every part of the human body so that the clothing ensemble can be customised precisely to the specificity of its application in a cold or hot climate (Bogdan and Zwolińska, 2012). In the following paragraphs, all of the applications of thermal manikins through known international standards shall be presented.

In 1995, ISO 9920 (ISO 9920:2007) was published and provided a list of the clothing ensembles with their respective insulation values as measured on manikins, for use in other standards assessing heat such as ISO 7933 (ISO 7933:2004), cold stress such as ISO 11079 (ISO 11079:2007) or thermal comfort in buildings such as ISO 7730 (ISO 7730:2005; Havenith and Nilsson, 2004). The database of clothing insulation values is provided in ISO 7933, ISO 9920 and ASHRAE 55 (ISO 9920:2007; ISO 7933:2004; ANSI/ASHRAE Standard 55-2017).

Assessment methods for prediction of the thermal characteristics of clothing, the thermal insulation and the water vapour resistance under reference conditions, including the effects of body movements and air velocity on the clothing's thermal insulation values and the water vapour resistance (ISO 9920:2007), are determined according *to* ISO 9920:2007 *Ergonomics of the Thermal Environment – Estimation of Thermal Insulation and Water-Vapour Resistance of a Clothing Ensemble.*

This standard describes the effects of clothing, such as adsorption of the water, the buffering or the tactile comfort. It also takes into account the influence of the rain and the snow on the thermal characteristics of the special protective clothing. The standard also deals with the separate insulation on the different parts of the body and the discomfort due to the asymmetry of the clothing ensemble. According to Parsons, this standard, along with *ISO 8996*, should be supplemented with the data on clothing from different parts of the world, the smart clothing and the dynamics of activity (Parsons, 2013).

The standard ISO 10551:1995 *Ergonomics of the Thermal Environment – Assessment of the Influence of the Thermal Environment Using Subjective Judgement Scales* proposes the set of specifications on direct expert assessment of the thermal comfort/discomfort expressed by the subjects (ISO 10551:1995). The persons are subjected to various degrees of thermal stress during periods spent in various workplace climatic conditions. The data provided by this assessment will be used to supplement the physical and the physiological methods of assessing thermal loads. This standard covers application of the judgement scales intended for providing reliable and comparative data on the subjective aspects of the thermal comfort felt in different working environments (ISO 10551:1995). This international standard is limited to the five scales of subjective judgement. The scales, presented in the standard, are divided into two types: personal scales and environmental scales. Those related to the personal thermal state may be perceptual, affective or preference scales. Those related to the environment may be acceptance and tolerance scales (Parsons, 2003).

Other standards from this series, such as ISO 7726 and ISO 7730, allow the experts to objectively evaluate thermal environments. These standards provide the methods for analytical determination of the indices, which are used to predict the average climatic conditions needed to obtain the thermal comfort or the degree of thermal stress felt by the persons in their working places. Most of the working areas show spatial heterogeneities with the local climatic differences and temporal fluctuations.

However, the subjective preferences of each subject differs. So it is important to take subjective judgements into consideration while designing the workplace and suggesting the most optimal working conditions. ISO 10551 implies this kind of specific analytical approach. At the beginning of each survey, one needs to construct and to choose the appropriate judgement scales. It is important to mention that the thermal environments which lend themselves to the application of subjective judgement scales relate to conditions which differ from a moderate degree or thermal neutrality. Under all testing procedures, one must take into account the appropriate deviation from the moderate working climate, so no health risks are set upon subjects. The data obtained by this standard is combined with other standards from this series of the ergonomics of the thermal environment (ISO 7243, 7726, 7730, 7933, 8996, 9886, 9920, 11079). There are a number of subjective judgement scales for the thermal environments. They differ whether the emphasis is placed on the specific time frame (present or past), the duration of the testing, whether the object of judgement is the subjective judgement of the environment or the consequences felt by the person while working in some environment, whether this is the perceptual or the affective judgement scale, whether the global or localised consequences of the organism are tested, whether the testing is instantaneous or extended over a period of time, etc. They also differ in the component tested (temperature, humidity, air movement, thermal state of the body, skin wetness, respiration), whether this is a permanent or temporary situation, due to conditions, etc. By repeatedly applying the same scales, the evolution with time of the thermal comfort or strain experienced in constant conditions may be assessed. An integrated judgement obtained over the whole time of exposure by appropriate computation of the data (e.g. overall mean) can also be obtained. In the case of steady climatic conditions, with sedentary working people, testing is applied after acclimatisation lasting 30 minutes. Persons submitted to repeated application of the same judgement scales should be informed beforehand. After responses to given questions in questionnaires are collected, the statistical analysis is performed.

Although subjective measurement techniques can be useful for measuring extreme environments, they should not be used as a primary measure of heat stress, cold stress and in health and safety since the ability of a person to make a 'rational' subjective judgement may be impaired (Parsons, 2003).

9.1.3 THE STANDARDS ASSESSING HEAT STRESS

Taylor et al. state that heat and cold stress are recognised as an occupational hazard within scientific literature, often attributed to compromised cognitive function. Heat stress-related cognitive decline is primarily mediated by a reduction in thermal comfort and/or changes in regional brain blood flow while both moderate and extreme reductions in ambient temperature may also have a negative effect on cognitive function (Taylor et al., 2016).

Outdoor conditions can present heat stress risks for people in hot climates. Heat stress can occur in unique situations, such as firefighting. Indoors, heat stress conditions occur in many workplaces, such as iron and steel foundries, glassmaking plants, bakeries, commercial kitchens, laundries, power plants, etc. Behavioural factors can amplify the risks, such as wearing impermeable clothing. Individual susceptibility to heat stress varies as a function of several physiological risk factors. Heat stress creates a progression of heat disorders with increasingly severe symptoms that can culminate in death (Havenith, 2005).

ISO 7243, ISO 7933 and ISO 9886 provide the current ISO system for the assessment of heat stress (Parsons, 1999, 2013). ISO 7243:2017 *Ergonomics of the Thermal*

Environment – Assessment of Heat Stress Using the WBGT (Wet Bulb Globe Temperature) Index, also known as BS EN 27243:2017, provides a method for monitoring the thermal conditions to which workers are exposed (ISO 7243:2017). The WBGT index value is used to predict the heat stress and thermal strain when working in hot environments together with additional factors of activity and clothing. ISO 7243 was first published in 1982 presenting the WBGT index in order to avoid the complicated determination procedure for the effective temperature index. The index combined temperature, humidity, radiation and wind into a single value (Parsons, 2006).

This standard presents the method for evaluating heat stress to which an individual is subjected in a hot environment. It applies to the evaluation of the mean effect of the heat on man during a representative period of some activity but it does not apply to the evaluation of heat stress suffered during very short periods nor to the evaluation of heat stresses close to the zones of comfort (Bethea and Parsons, 2002; Parsons, 2013). Also, insufficient guidance on the impact of clothing on heat stress prevention is mentioned (Parsons, 2006).

The methods for evaluating health risks, while exposed to hot environments, on the basis of variation from heat-balance conditions between the human body and the environment, are covered by ISO 7933:2004 *Ergonomics of the Thermal Environment – Analytical Determination and Interpretation of Heat Stress Using Calculation of Predicted Heat Strain* (ISO 7933:2004). This standard provides limiting values for work in hot environments based upon heat storage and dehydration. The duration-limited exposures for dehydration are calculated upon a maximum water loss. Warning level duration-limited exposures are calculated to around 4% and for a danger level of around 6% of body weight for non-acclimatised and acclimatised people, respectively (Parsons, 2003).

The standard was published for the first time in 1989 (Malchaire et al., 2002). This method is acceptable in cases where classical clothing is worn (personal protective and functional clothing are not addressed by this standard), and it describes the method for predicting the sweat rate and the internal core temperature that the human body will develop in response to the hot environmental conditions. This is the means by which to evaluate the thermal stress felt by the subject and physiological strain. It is also suitable for prediction of the maximum allowable exposure times in hot environments. This standard, together with ISO 9920:2007, is the foundation for the determination of the evaporative resistance of the clothing ensemble, although Havenith et al. debated the adequacy of the proposed relations presented in those standards when the wearer is moving or is exposed to wind (Havenith et al., 1999). Holmér et al. also proposed new equations for determination of the reduction of the total insulation values obtained under static, still wind conditions as a consequence of wind and walking effects. The former calculation of convective and radiative heat losses in ISO 7933, from 1989 underestimated the values due to insufficient consideration of the effects of body motion and wind on clothing heat transfer (Holmér et al., 1999). This highlights the importance of continual revisions and of developing standards which best reflect how materials may behave in suitably representative environments.

Havenith et al. corrected the calculation of clothing insulation ($I_{T\,dynamic}$), published an extended table for the vapour permeability index (i_m) and derived a new empirical model improving the effects of wind and movement on the permeability index (Havenith et al., 1999). This research study proposed corrections for dynamic clothing vapour resistance due to air movement and walking speed by testing the two main methods, one proposed by ISO 7933:1989 and other by ISO 9920:1995, together with the four alternative approaches (Havenith et al., 1999).

According to the older version of ISO 7933:1989, the total water vapour resistance (R_{eT}) of the clothing ensemble can be estimated by means of the reduction factors for latent heat exchange (F_{pcl}) and the evaporative heat transfer coefficient (h_e) (Havenith et al., 1999):

$$R_{eT} = \frac{1}{h_e \cdot F_{pcl}}, h_e = 16.7 \cdot h_c \quad \left[\text{kPa}\,\text{m}^2/\text{W}\right] \quad (9.1)$$

But, as the intrinsic clothing insulation (I_{cl}) is a constant, the reduction factor for latent heat exchange (F_{pcl}) is dependent on the convective heat transfer coefficient (h_c), which changes when the person starts moving or when the wind is present (Havenith et al., 1999):

$$I_{cl} = 1/1 + 2.22\,h_c \times \left\langle I_{cl} - \left\{\left[1 - \left(1/f_{cl}\right)\right]/\left(h_c + h_r\right)\right\}\right\rangle \quad \left[\text{m}^2\text{K}/\text{W}\right] \quad (9.2)$$

Where F_{pcl} is permeation efficiency (the ratio of the actual evaporative heat loss to that of a nude body at the same conditions, including an adjustment for the increase in surface area due to the clothing) and h_r is the radiative heat transfer coefficient (W/m²K).

When taking a constant intrinsic clothing insulation approach, an increase in the wind speed resulted in a decreasing reduction factor for the latent heat exchange since the vapour resistance of the boundary air layer (R_a) reduces much more by wind than the total water vapour resistance (R_{eT}) (Havenith et al., 1999):

$$F_{pcl} = R_a / R_{eT} \quad (9.3)$$

According to former ISO 9920:1995, there is also a second main method of calculating the total water vapour resistance (R_{eT}) of the clothing ensemble. It can be estimated in the relation to the dry heat resistance (based on the total insulation, I_T) and by means of the vapour permeability index (i_m) and the Lewis relation (L):

$$R_{eT} = I_T / (i_m \cdot L) \quad \left[\text{kPa}\,\text{m}^2/\text{W}\right] \quad (9.4)$$

But since ISO 9920 provides assumptive data on the vapour permeability index for the permeable clothing regardless of the number of the layers, those values have been observed and found to be lower than those found in *ISO 9920*. New corrections have been proposed for the dynamic permeability index of clothing ($i_{m,\,dynamic}$), the total dynamic clothing insulation ($I_{T,\,dynamic}$) and the dynamic clothing water vapour resistance ($R_{T,\,dynamic}$) (Havenith et al., 1999). Alternative approaches have also been examined.

Of all the six reviewed methods, the authors concluded that the best method is to use the corrected value of the dynamic clothing insulation ($I_{Tdynamic}$) and the

Overview of the Most Significant Standards on Thermal Comfort

corrected value of the dynamic vapour permeability index ($i_{m,dynamic}$) to determine the dynamic vapour resistance ($R_{Tdynamic}$):

$$R_{T\,dynamic} = I_{T\,dynamic} / (i_{m\,dynamic} \cdot L) \quad [kPa\,m^2/W] \quad (9.5)$$

According to Parsons et al. (Parsons et al., 1999) when a clothed person is exposed to wind and activity, there is a potentially significant limitation in the simplified clothing model presented in ISO 7933. The heat and the mass transfer can take place between the microclimate (within the clothing and next to the skin surface) and the external environment, causing the intrinsic clothing insulation reduction due to the effects of the wind and the human movement. The effective decreased clothing insulation and the increased vapour permeability will allow greater heat transfer by convection and the evaporation.

According to a HSE report, the methods described in ISO 7933 are often not sufficient to meet the varied realities of the industrial heat stress (Bethea and Parsons, 2002).

Mehnert et al. (Mehnert et al., 2000) and Malchaire, J. et al. validated the former versions of *ISO 7933:1989* (Malchaire et al., 1999, 2000a, 2000b). Both Mehnert et al. and Malchaire, J. et al. published the study under the research project 'Assessment of the risk of heat disorders encountered during work in hot conditions' for the prediction of the mean skin temperature in warm and the hot environments (Malchaire et al., 2002). The limit criteria for estimating the acceptable exposure times in hot working conditions according to *ISO 7933:1989* were revised. Malchaire et al. proposed a revision of the required sweat rate model to a new model named the Predicted Heat Strain (PHS) index (Malchaire et al., 1999, 2000a, 2000b).

The model used for calculating the required sweat rate index (SW_{req}) and for predicting the mean skin temperature, in the former version of ISO 7933:1989, was criticised for not being valid in conditions with high radiation and high humidity. Malchaire's study proposed a new model called the Predicted Heat Strain Model which tends to predict the sweat rate (SW_p) and the rectal temperature (t_{re}) minute-per-minute as a function of the working conditions (Malchaire et al., 2002). They gathered the larger database containing the mean skin temperature values for validation of the PHS model and new mean skin temperature (t_{sk}) prediction for nude and clothed subjects was derived for warm and hot environments:

For the nude subjects, a new mean skin calculation was proposed:

$$t_{sk} = 7.19 + 0.064 \cdot t_a + 0.061 \cdot t_r + 0.198 \cdot p_a - 0.348 \cdot v_a + 0.616 \cdot t_{re} \quad (9.6)$$

For the clothed subjects, a new mean skin calculation was also proposed:

$$t_{sk} = 12.17 + 0.02 \cdot t_a + 0.044 \cdot t_r + 0.194 \cdot p_a - 0.253 \cdot v_a + 0.00297 \cdot M + 0.513 \cdot t_{re} \quad (9.7)$$

Where v_a is air velocity (m/s), t_r is mean radiant temperature (°C), t_{re} is rectal temperature (°C), t_a is air temperature (°C), p_a is water vapour pressure (kPa), and M is the metabolic rate of the body (W/m²).

The proposed model for the mean skin temperature calculation extends the validity to conditions with a high radiant heat load or high humidity. Since then, the second and third editions of ISO 7933:2004 have been modified in accordance the with PHS model (ISO 7933:2004).

Although, the impact of heat stress is covered by ISO 7243:2017 and ISO 7933:2004, *only ISO 9886:2004* provides the basis of the physiological measurements to indicate the thermal strain and can be applied to monitor the individual response of the workers, while ISO 7243:2004 provides a simple monitoring method for assessing heat stress and ISO 7933:2017 provides an analytical approach. Neither of them is valid while wearing protective clothing or during the short or rapidly changing exposures (Holmér, 2006; Parsons, 2013).

The standard ISO 9886:2004 *Ergonomics – Evaluation of Thermal Strain by Physiological Measurements* describes measurements and interpretations of the physiological parameters of the human body, which are the basis for estimating strain possibly sustained by a subject during exposure in a specific environment or while performing certain activities (ISO 9886:2004). The standard describes methods for measuring and interpreting the following physiological parameters such as the body core temperature, skin temperatures, heart rate and body mass loss. The body core temperature could be approximated on the basis of oesophageal temperature, rectal temperature, intra-abdominal temperature, oral temperature, tympanic temperature, auditory canal temperature and urine temperature. Depending on the technique used, the temperature measurement can reflect the mean temperature of the body mass, or the temperature of the blood irrigating the brain and therefore influencing the thermoregulation centres in the hypothalamus.

The measurements of mean skin temperature are performed at a different location on the body, because skin temperatures aren't homogeneous at different parts of the body, using a temperature transducer with a precision of ±0.1 degree in a range from 298.15 to 313.15 K (25–40°C). This skin's heterogeneity is primarily the consequence of the ambient conditions. One can measure the local skin temperature (t_{sk}) measured at a specific point on the body surface, and the mean skin temperature on the entire surface of the body. The skin temperature is influenced by the thermal exchanges by conduction, convection, radiation and evaporation at the surface of the skin, the variations of skin blood flow and the temperature of the arterial blood reaching the particular part of the body.

Conductive devices fixed to the surface of the skin were employed in the first skin temperature assessment. The conductive devices measure the transfer of heat energy into the device through direct contact with the object of interest. Currently, there are three main types of conductive temperature-measuring sensors, thermocouples, thermistors and telemetry sensors. Thermocouples have a fast response and accuracy over very wide temperature ranges. They measure temperature proportional to the voltage between two dissimilar metals when in the presence of a temperature gradient. Thermistors are highly thermally sensitive and wired sensors from a semi-conductive metal with an inverse relationship between temperature and the electrical resistance within the device, which show fast response rates and high accuracy when measuring temperatures within the human physiological range. Both thermocouples and thermistors require an associated data logger to control sampling rates and store recorded data. The wireless thermochron sensors, also called wireless telemetry systems, are an autonomous encapsulated semiconductor computer chip. There is a limitation when using the conductive contact temperature sensors (thermocouples, thermistors and telemetry sensors), such as wire entanglement, loss of sensor contact during exercise, the formation of a microenvironment around the sensor due to the fixation method and relatively small spot measurements.

Nowadays, new infrared non-contact devices are increasingly utilised for skin temperature measurements (infrared thermometers and infrared thermal imaging cameras). Similar to conductive devices, infrared thermometers measure the average temperature over a small area (spot measurement). However, the infrared devices may not be suitable for monitoring the mean skin temperature in the presence of or following heat stress induced by metabolic and environmental changes. During exercise in the heat and subsequent recovery, infrared devices may not be sufficiently accurate to determine the mean skin temperature. For example, during the rest period, the infrared devices significantly overestimated the mean skin temperature compared to the thermistor (Bach et al., 2015a, 2015b; James et al., 2014).

To determine the thermal conductivity and the thermal diffusivity of biomaterials, frequently thermal probe techniques are used, most often a thermistor bead either as a heat source or a temperature sensor inserted within the tissue. Thermal probes have a miniature thermistor at the tip of a plastic catheter, and the volume of tissue measured depends on the surface area of the thermistor, while the electrical power and the resulting temperature rise are measured. The important factors for the thermal probe are thermal contact and transducer sensitivity. The shape of the probe should be chosen to minimise trauma during insertion. Electrical power is delivered to a spherical thermistor within the tissue, which is assumed to be homogeneous within the millilitre surrounding the probe. The thermistor measures the combination of conduction and latent heat when the initial tissue temperature is just below freezing point. When the tissue is perfused by blood, the thermistor measures the heat removal by conduction and heat transfer due to blood flow. However, there are pitfalls with such invasive techniques that have to be addressed when using thermal tissue measurements with micro-probes inserted within the tissue. A fluid pool may form around the probe because of the mechanical trauma and tissue damage may occur, causing the presence of a pool of blood and other fluids that significantly alters the results. Second, the probe measures a non-uniform spatial average of the heterogeneous tissue properties. The limitation of most techniques is that the intrinsic tissue thermal conductivity must be known in order to accurately measure perfusion (Diller et al., 2000).

The non-invasive sensors are currently the most promising innovations to monitor physiological functions, which offer painless applications and comfort. These sensors can be applied either in contact with the body or near to it (wearable) or can be embedded in surroundings. These are non-invasive micro-sensors, biomedical sensors, implants and disposable biochips for blood analysis and clinical diagnosis (Lymberis and Olsson, 2003).

The assessment of the thermal strain from the heart rate is based on measuring the beats per minute over a time interval. The increase in heart rate is usually related to an increase in the core temperature and thermal stress due to environmental conditions. The increase in heart rate for an increase of 1 degree in the body core temperature is called the thermal cardiac reactivity and is expressed in heartbeats per minute and per degree Celsius $[\text{bmp/min} \cdot {}^\circ\text{C}]$.

The assessment of the physiological strain on the basis of the body mass loss due to sweating is determined based on the gross body mass loss of a person during a given time interval. When assessing the physiological strain on the basis of the body mass loss due to sweating, one will determine the sweat loss and the net water balance of the body. In warm and hot environments, sweat loss can be considered as an index of physiological strain, including not only the sweat that evaporates at the surface of the skin but also the fraction dripping from the body surface or accumulating in the clothing. The net water balance is important when considering the risk of dehydration. Regular intake of small volumes of water are able to balance for about 75% of the water loss.

9.1.4 THE STANDARDS ASSESSING COLD STRESS

The exposure to cold environments comprises a significant hazard since cold results in different types of cold stress. The types of cold stress are whole-body cooling, extremity cooling, skin convective cooling, skin conductive cooling and respiratory cooling. The associated environmental climatic stress factors for whole-body cooling and extremity cooling are air temperature, mean radiant temperature, air velocity, relative humidity, activity level and clothing. The environmental stress factors for convectional skin cooling are the air temperature and the air velocity, while the surface temperature and clothing are the main ambient stress factors when dealing with conductional skin cooling. When speaking of respiratory cooling, air temperature and activity level are the main concern (Holmér, 1998). Many international standards are deployed to analyse the stress, the strain and the risks imposed on the human body.

Cold stress is developed when some, or several, of the basic avenues of heat loss (radiation, evaporation, conduction and convection) exceed the heat production of human tissues. The available methods of assessing cold stress analyse one or several forms of the avenues of heat loss, or they quantify the signs of cold stress in the human body. There are several indices and standards to quantify cold stress (Havenith, 2005).

The low ambient temperatures are the primary parameter which affects the human body, endangering human health. The standard ISO 11079:2007 *Ergonomics of the Thermal Environment – Determination and Interpretation of Cold Stress When Using Required Clothing Insulation (IREQ) and Local Cooling Effects* (ISO 11079:2007) describes the impact of the cold felt under windy conditions usually in cold climatic conditions by the *wind chill index* (WCI). There are two cold stress indices, the WCI and the required clothing insulation (IREQ) index. The WCI is often regarded as a good indicator of local cooling of the hands, feet, face and exposed skin, which allows the effects of air temperature and wind to be combined to predict the effects on clothed people. The IREQ index is regarded as a whole-body cold stress index which accounts for the clothing insulation required to be worn to maintain comfort ($IREQ_{neutral}$) or in just acceptable conditions ($IREQ_{min}$) where a person will become cold but not unacceptably cold. In general, the cold stress index integrates the effects of relevant factors into a single value which can be related to cold strain and therefore can be used to quantify the extent of the cold stress on the workers and to provide limits for safe and comfortable working as well as guidance on how to design an environment and working practices that might alleviate cold strain (Parsons, 1998). ISO 11079:2007 deals with all types of cold stress, whether this is the whole-body cooling, local cooling, extremity cooling, skin cooling by wind, skin cooling by contact or airway cooling (Holmér, 2009).

$IREQ_{neutral}$ is an analytical index for cold stress that integrates the effects of the air and the mean radiant temperature, the humidity, the velocity and the metabolic rate. It defines the insulation required to maintain the thermal equilibrium at a normal level with no or minimum cooling of the body (Griefahn, 1998). However, the serial method provides a higher insulation value for a clothing ensemble compared to the parallel method both for static and walking conditions (Kuklane et al., 2007).

There are also to national standards regulating working practices in the cold, one being the DIN 33403-5:1994 *Klima am Arbeitsplatz und in der Arbeitsumgebung – Teil 5: Ergonomische Gestaltung von Kältearbeitsplätzen* (DIN 33403-5:1997) and another being BS 7915:1998 *Ergonomics of the Thermal Environment – Guide to Design and Evaluation of Working Practices for Cold Indoor Environments* (BS 7915:1998). DIN 33403-5 is based upon the IREQ index but doesn't apply to outdoor work. It defines the cold environments as those from 288.15 to 223.15 K (15 to −50°C). DIN 33403-5 provides an assessment method, the tables of minimum clothing insulation required, general guidance on working practices and ergonomics measures for reducing cold strain. BS 7915:1998 gives guidance on ways in which cold stress or discomfort in cold indoor environments can be evaluated and cold strain reduced. Cold environments are defined as those with an air temperature of less than 285.5 K (12°C). It provides an assessment method, the description of the human responses to the cold together with the influence of different working practices, as the WCI and the IREQ index (Parsons, 1998).

Humans have the ability of both unconscious and conscious adaptation to climatic conditions. Thermoregulatory control mechanisms can be unconscious and conscious. The unconscious mechanisms are described through negative feedback, the hypothalamus and the hormones, while conscious thermoregulatory control mechanisms include behavioural changes such as moving closer to the heat source when it is cold, dressing in extra layers of thicker clothing or removing extra layers of clothing when it's hot. Clothing is the main conscious factor in maintaining thermal neutrality and acquiring a state of thermal comfort under the dynamic climatic conditions and due to the reactions inside the human body. Therefore, it is important to know the insulation values of the clothing garments, and the ensembles and requirements then imposed. Heat exchange in the human organism has its origin above all in the heat-balance equation. Thus, it is very important to include the heat exchange at the surface of the skin and the effect of clothing as the main barrier blocking the heat release from the human body to the environment. This standard is compatible to other standards (ISO 7726, 8996, 9237, 9920, 13731, 13732-3, 15831 and 511), which set the requirements while measuring human reactions in warm environments, parameters assessment and while describing thermal environments, the fabric type and clothing ensembles for protection against the cold.

The physical properties of the textile materials which contribute to physiological comfort involve a complex combination of the heat and the mass transfer. Each may occur separately or simultaneously. They are time-dependent and may be considered in steady-state or transient conditions. The measuring methods for textile materials are described through the standard ISO 11092:2014 (former BS EN 31092:1993) *Physiological Effects – Measurement of Thermal and Water-Vapour Resistance Under Steady-State Conditions (Sweating Guarded-Hotplate Test)* (ISO 11092:2014).

Evaluation of clothing fibres, fabrics, membranes and the like should not be done with manikins. Their five key parameters are too often masked by such factors as drape, fit, specific contact area and weight to glean appropriate information from manikin studies. These material properties are 1) fabric insulation; 2) fabric moisture permeability; 3) wicking characteristics; 4) water uptake/holding characteristics; and 5) drying time. The first studies on the heated guarded-flatplate apparatus for measuring the insulation properties of new materials demonstrated that clothing insulation was a linear function of the increasing circumference of the layers of clothing, and the air layers trapped between them, with practically negligible influence from any specific aspect of the materials or their fibres except their thickness (Goldman, 2006).

Thermal resistance is the complex result of the combination of radiant, conductive and convective heat transfer, and its value depends on the contribution of each component to the total heat transfer. Although it is the intrinsic property of the textile material, its measured value may change with the conditions of testing due to the interaction of the parameters such as the radiant heat transfer with the surroundings. There are several methods, which may be used to measure the heat and the moisture properties of textiles, each of them being associated with a certain property and relying on a certain assumption for its interpretation. The sweating guarded-hotplate, which is often called the skin model, is described in this standard and is intended to simulate the heat and mass transfer processes, which occur next to the human skin.

Measurements involving one or both processes may be carried out either separately or simultaneously using a variety of environmental conditions. These conditions involve combinations of the temperature, the relative humidity, the airspeed, either in the liquid or the gaseous phase. Hence, transport properties measured with this apparatus can be made to simulate different wear and environment conditions in both transient and steady states. This standard covers only steady-state conditions, while the application area is defined through specifying the thermal and the water vapour resistance under steady-state conditions, of, e.g. the fabrics, films, coatings, foams and leather, including the multilayer assemblies, for use in clothing, quilts, sleeping bags, upholstery and similar textile or textile-like products. The application of this measurement technique is restricted to maximum thermal resistance and water vapour resistance which depend on the dimensions and construction of the apparatus used (e.g. $R_{ct}=2$ m^{2}K/W and $R_{et}=700$ m^{2}Pa/W respectively, for the minimum specifications of the equipment). The test conditions are not intended to represent the specific comfort situations, and the performance specifications in relation to the physiological comfort are not stated (ISO 11092:2014).

Although clothing characteristics for the calculation of the sensible heat exchange are described in international standards such as ISO 7730:2005, ISO 7933:2004 and

Overview of the Most Significant Standards on Thermal Comfort 205

ISO 9920:2007, there are a few European standards describing the specification and the requirements of the protective clothing garments and ensembles.

The protective clothing ensembles for the protection of the human body against cold are described by standard BS EN 342:2004 (BS EN 342:2004). The standards for describing cold-protective clothing are shown in Figure 9.2. In comparison, there is a vast number of standards for describing the requirements and measuring methods for protective clothing against heat and flames, see Figure 9.3.

Single clothing garments for protection from the effects of cool environments are described by BS EN 14058:2004 (BS EN 14058:2004). In BS EN 14058:2004 *Protective Clothing – Garment for Protection Against Cool Environments*, the requirements on single clothing garments, which should be achieved, are stated to protect against local body cooling. The thermal stress sustained due to the cold is observed from the point of entire body cooling, called *general cooling* and partial body cooling of a certain area of the body is called *local cooling*.

In the standard BS EN 14058:2004, the focus is on describing the consequences of local body cooling and its impact on obtaining the feeling of thermal comfort. With local body cooling, the susceptible parts of the body are the extremities such as the hands, feet and face. The single garments which cover a part of the body are described, such as the undershirt with long sleeves, the long underpants, the

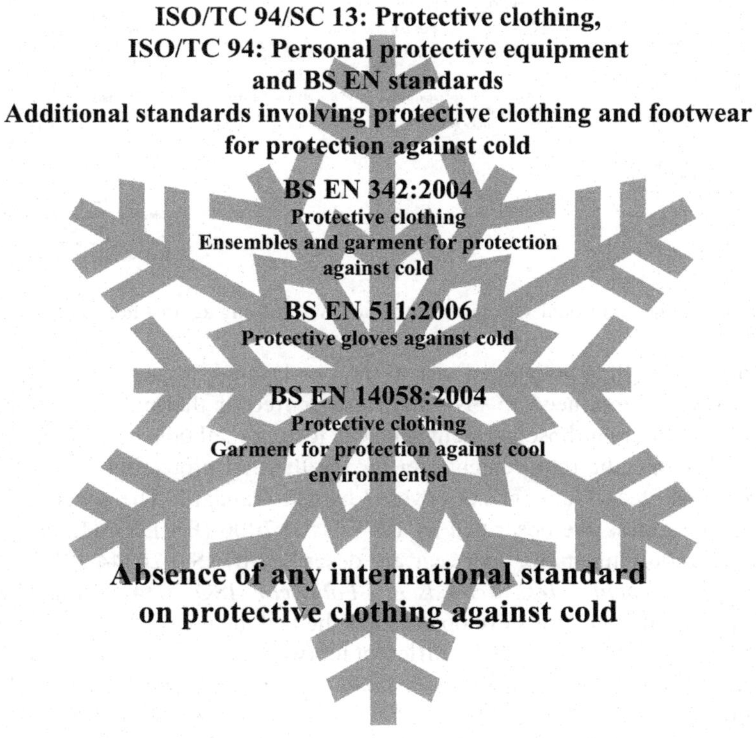

**ISO/TC 94/SC 13: Protective clothing,
ISO/TC 94: Personal protective equipment
and BS EN standards
Additional standards involving protective clothing and footwear
for protection against cold**

BS EN 342:2004
Protective clothing
Ensembles and garment for protection
against cold

BS EN 511:2006
Protective gloves against cold

BS EN 14058:2004
Protective clothing
Garment for protection against cool
environmentsd

**Absence of any international standard
on protective clothing against cold**

FIGURE 9.2 The additional standards for cold-protective clothing.

ISO/TC 94/SC 13: Protective clothing, ISO/TC 94: Personal protective equipment and BS EN standards
Additional standards involving protective clothing and footwear for protection against heat and flame

BS EN 407:2004
Protective gloves against thermal risks (heat and/or fire)

ISO/TR 2801:2007
Clothing for protection against heat and flame
General recommendations for selection, care and use of protective clothing

ISO 6942:2002
Protective clothing
Protection against heat and fire
Method of test: Evaluation of materials and material assemblies when exposed to a source of radiant heat

ISO 9151:2016
Protective clothing against heat and flame
Determination of heat transmission on exposure to flame

ISO 11612:2015
Protective clothing
Clothing to protect against heat and flame
Minimum performance requirements

ISO 12127-1:2015
Clothing for protection against heat and flame
Determination of contact heat transmission through protective clothing or constituent materials
Part 1: Contact heat produced by heating cylinder

ISO 12127-2:2007
Clothing for protection against heat and flame
Determination of contact heat transmission through protective clothing or constituent materials

ISO 13506-1:2017
Protective clothing against heat and flame
Part 1: Test method for complete garments
Measurement of transferred energy using an instrumented manikin

ISO 14116:2015
Protective clothing
Protection against flame
Limited flame spread materials, material assemblies and clothing

ISO 14460:1999
Protective clothing for automobile racing drivers
Protection against heat and flame
Performance requirements and test methods

ISO 15025:2016
Protective clothing
Protection against flame
Method of test for limited flame spread

ISO 17492:2003
Clothing for protection against heat and flame
Determination of heat transmission on exposure to both flame and radiant heat

ISO 17493:2016
Clothing and equipment for protection against heat
Test method for convective heat resistance using a hot air circulating oven

ISO 11613:2017
Protective clothing for firefighter's who are engaged in support activities associated with structural fire fighting
Laboratory test methods and performance

ISO 11999-1:2015
PPE for firefighters
Test methods and requirements for PPE used by firefighters who are at risk of exposure to high levels of heat and/or flame while fighting fires occurring in structures
Part 1: General

ISO/TS 11999-2:2015
PPE for firefighters
Test methods and requirements for PPE used by firefighters who are at risk of exposure to high levels of heat and/or flame while fighting fires occurring in structures
Part 2: Compatibility

ISO 11999-3:2015
PPE for firefighters
Test methods and requirements for PPE used by firefighters who are at risk of exposure to high levels of heat and/or flame while fighting fires occurring in structures

ISO 11999-4:2015
PPE for firefighters
Test methods and requirements for PPE used by firefighters who are at risk of exposure to high levels of heat and/or flame while fighting fires occurring in structures
Part 4: Gloves

ISO 11999-6:2016
PPE for firefighters
Test methods and requirements for PPE used by firefighters who are at risk of exposure to high levels of heat and/or flame while fighting fires occurring in structures
Part 6: Footwear

ISO 15383:2001
Protective gloves for firefighters
Laboratory test methods and performance requirements

ISO 15384:2018
Protective clothing for firefighters
Laboratory test methods and performance requirements for wildland firefighting clothing

ISO 15538:2001
Protective clothing for firefighters
Laboratory test methods and performance requirements for protective clothing with a reflective outer surface

ISO 16073:2011
Wildland firefighting personal protective equipment
Requirements and test method

ISO 18640-1:2018
Protective clothing for firefighters
Physiological impact
Part 1: Measurement of coupled heat and moisture transfer with the sweating torso

ISO 18640-2:2018
Protective clothing for firefighters
Physiological impact
Part 2: Determination of physiological heat load caused by protective clothing worn by firefighters

ISO 17493:2016
Clothing and equipment for protection against heat
Test method for convective heat resistance using a hot air circulating oven

FIGURE 9.3 The additional standards for protective clothing against heat and flames.

up to knee socks, the one-layered jacket, the one-layered trousers, the vest and the coat. These garments need to give a certain degree of thermal protection. Both requirements set upon those garments and the testing methods are described in the standard. The specific requirements set for the thermal protection of the extremities (e.g. hands, feet, face) are not covered by this standard. The specific requirements for the gloves are described by BS EN 511:2006 (BS EN 511:2006), while the specific requirements for footwear are described by ISO 20344:2011 *Personal Protective Equipment – Test Methods For Footwear* (ISO 20344:2011). There are also international standards developed under the technical committee ISO/TC 216, which describe the test methods for different footwear parts such as ISO 17699:2003 *Footwear – Test Methods for Uppers And Lining – Water Vapour Permeability and Absorption* (ISO 17699:2003), ISO 17702:2003 *Footwear – Test Methods for Uppers – Water Resistance* (ISO 17702:2003), ISO 17705:2003 *Footwear – Test Methods for Uppers, Lining and Insocks – Thermal Insulation* (ISO 17705:2003), ISO

Overview of the Most Significant Standards on Thermal Comfort

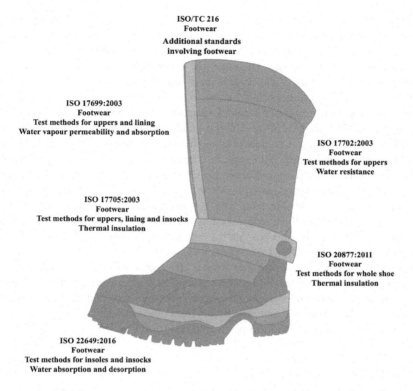

FIGURE 9.4 The additional international standards which describe the measurement methods for footwear thermal and moisture properties.

22649:2016 *Footwear – Test Methods for Insoles and Insocks – Water Absorption and Desorption* (ISO 22649:2016) and ISO 20877:2011 *Footwear – Test Methods for Whole Shoe – Thermal Insulation* (ISO 20877:2011), see Figure 9.4.

The thermal protection of the head is also described in the above-mentioned standard BS EN 342:2004 *Protective Clothing – Ensembles and Garments for Protection Against Cold* (BS EN 342:2004). However, this standard describes only one type of protective garments for head protection against the cold – the *balaclava*. The level of thermal protection, which should be achieved by these garments, depends primarily on the application of these garments. If their usage is mainly for protection while working under low temperatures and for long exposures, the level of thermal protection of specific clothing garments should be precisely assessed. These garments could be used for work in both indoor and outdoor environments. In these cases, the standard is referred to moderate low temperatures, which don't deviate much from the optimum working temperatures, but could affect the working capabilities during prolonged exposures (e.g. while working indoor where the lower temperatures are required like for the cold storage in the food processing industry).

BS EN 342:2004 *Protective Clothing Ensembles and Garment for Protection Against Cold* (BS EN 342:2004), defines the requirements and the testing methods for protective clothing, especially the clothing ensembles and afterwards the singles clothing garments for protection against the cold. Thermal insulation is one of the most important properties which is tested by the thermal manikin apparatus. The convective heat losses are increased by the effect of wind. Therefore, the water vapour permeability, apropos the water vapour permeability of the outer layer of the clothing ensemble and the thermal insulation of clothing, are important factors to be taken into account in relation to the protection of the wearer against the cold. Except for the thermal insulation and the water vapour permeability, the protective clothing should also have a defined ability for moisture absorption. Sweating is an unavoidable physiological process but is also extremely dangerous during prolonged cold exposure. The moisture absorption will progressively reduce the garment's thermal insulation and the degree of thermal protection. Thereby for protective clothing, whose main purpose is thermal protection under cold conditions, it is important to select flexible, adjustable garments which will adapt to the body contours rather than to select the maximal insulation level. When designing such protective garments one should take into account the ventilation openings, so that excessive heat and moisture management and the spontaneous cooling of the body can be ensured. Passive diffusion of excessive moisture is not as efficient as active ventilation through the foreseen ventilation openings. When the environment temperatures are below 283.15 K (10°C), the water vapour released into the environment is weakened due to condensation and freezing, which takes place inside the structure of the material.

Risikko et al. summarised the performance criteria for cold-protective clothing according to European/ISO standards, Table 9.2 (Risikko et al., 1998).

According to ISO 11079, for active people, clothing insulation can be reduced by 50% in the cold due to ventilation (Parsons, 2003).

The test procedure used to measure thermal insulation of a clothing ensemble by means of a thermal manikin is described through standard ISO 15831:2004 *Clothing – Physiological Effects – Measurement of Thermal Insulation by Means of a Thermal Manikin* (ISO 15831:2004). Testing is performed to determine the efficiency of the thermal insulation of the clothing, which the wearer will use in relatively calm environments. The thermal insulation of clothing can be used to determine the physiological effect of the clothing on the wearer in a specific climate and under different activity scenarios. There are a number of thermal manikin constructions. The measurement could be performed with the manikin stationary or with the extremities moving, simulating a human walk.

ISO 15831:2004 specifies the following terms:

- *Clothing ensemble* is a group of garments worn together on the body at the same time.
- *Thermal insulation of clothing* is a temperature difference between the wearer's skin surface and the ambient atmosphere divided by the resulting dry heat flow per unit area in the direction of the temperature gradient where the dry heat flow consists of conductive, convective and radiant components. Depending on the end use of the clothing, different thermal insulation values can apply.
- *Total thermal insulation of clothing*, I_T, is total thermal insulation from the skin to the ambient atmosphere, including the clothing and the boundary air layer, under defined conditions measured with the stationary manikin.

TABLE 9.2
The Objectives for the Protective and Functional Properties that a Good Outdoor Work Garment Should Reach in Various Tasks and Different Weather Conditions

Parameter	Limit	Basis	Method
Thermal insulation for long-term cold exposure	≥3 clo	IREQ from ISO 11079:2007	BS EN 342:2017
Thermal insulation for short-term cold exposure	≥2.5 clo	IREQ from ISO 11079:2007	BS EN 342:2017
Thermal insulation for gloves	2.5–3 clo	Frostbite	BS EN 511:2006
Air permeability for wind, rest (100 W/m^2)	<20 l/m^{2}s	Preventing of cooling	ISO 9237:1995
Air permeability for heavy work (>300 W/m^2)	20–150 l/m^{2}s	evaporation	ISO 9237:1995
Ventilation for heavy work (>300 W/m^2)	>300 l/min	–	–
Water vapour permeability for cold weather clothing	≤13 m^{2}Pa/W	Evaporation	ISO 11092:2014 (BS EN 31092:1993)
Water vapour permeability for cold/foul weather clothing	≤20 m^{2}Pa/W	Evaporation	ISO 11092:2014 (BS EN 31092:1993)
Resistance to water penetration under light rain	≥2200 Pa	Protection against moisture	ISO 811:2018
Resistance to water penetration under heavy rain	≥13000 Pa	Protection against moisture	ISO 811:2018

(ISO 11092:2014; BS EN 342:2004; BS EN 511:2006; Risikko et al., 1998; ISO 811:2018; ISO 9237:1995)

Resultant total thermal insulation of clothing, I_{Tr} is the total thermal insulation from the skin to the ambient atmosphere, including the clothing and the boundary air layer, under defined conditions measured with the manikin moving its legs and arms.

A comprehensive study has been conducted and published by Anttonen et al. in 2004 on the precision of thermal manikin measurements (Anttonen et al., 2004). Due to variation in the construction of the manikins and in ambient conditions, it is important to evaluate repeatability, reproducibility and between-laboratory variance. Also, discussion of the methods for calculating insulation and their relevance is important. The study showed that none of the manikin measurement options (walking/parallel, walking/serial, standing/parallel, standing/serial) gave results, which in all conditions would be unambiguously reliable for predicting thermal protective properties. The influence of ventilation was assessed by measurement using a walking manikin. However, it would be most informative to have both standing and walking values marked in cold-protective clothing. The influence of the heating regulation system on the insulation values was minimal, so both comfort and normal control systems can be used. The effect of ambient conditions was critical only in the case of air velocity, which should be defined by speed and the direction of airflow.

Air temperature should be cool enough to keep the control unit working actively. Air humidity was not a critical parameter. All of the parameters were evaluated by the criteria to keep the deviation in the insulation results due to individual ambient parameter to less than 5%. The criteria are obviously stricter when calculating the resultant maximum error, caused by all the different parameters together. That is why instructions should be given in the manikin standard describing the ambient thermal parameters. The possible influence of washing and wearing on the cold-protective properties is mentioned in the introductory paragraph of *ENV 342*. However, thermal insulation tests are performed on new garments. If customers buy cold-protective systems for particular conditions, it is important that the protective properties remain throughout the entire period of usage. Therefore, at least an optional additional test after a number of washing cycles is recommended (Anttonen et al., 2004).

Other significant ISO standards in the field of the ergonomics of the thermal environment are presented in Table 9.3.

9.2 THE MOST SIGNIFICANT ASTM STANDARDS ON THERMAL COMFORT

The impact of the overall thermal transmission coefficient of textile materials (or thermal transmittance) is determined by ASTM D 1518-85 under the title *Standard Test Method* for Thermal Transmittance of Textile Materials (ASTM D1518-85, 2003), which was renamed ASTM D1518-14 *Standard Test Method for Thermal Resistance of Batting Systems Using a Hot Plate* in 2014 (ASTM D1518-14, 2014), Table 9.4. In the 2003 standard, the procedures described were used to identify the impact of the overall thermal transmission coefficient, which is the outcome of the combined action of conduction, convection and radiation for dry specimens of textile fabrics, battings and other materials. The overall thermal transmission coefficient is used to measure the time rate of heat transfer from a warm, dry, constant-temperature, horizontal flatplate up through a layer of the test material to a relatively calm, cool atmosphere (ASTM D1518-14, 2014). In the 2014 standard, the thermal resistance of a batting or batting/fabric system is measured. Both ISO 11092:2014 (ISO 11092:2014) and ASTM D1518-14 (ASTM D1518-14, 2014) standards are important for determining the fabric's thermal resistance and their further application in fabricating cold weather protective clothing, sleeping bags and bedding systems. The thermal interchange between man and his environment is too complicated a subject, and measured fabrics' thermal insulation values can only indicate the relative merits of a particular material. In addition, further measurements of the overall and intrinsic clothing insulation under standard uniform and different non-uniform environmental conditions should be performed in order to get a bigger picture on the impact of fabric layers as a barrier to thermal interchange between the human body and the surrounding environment.

The ASTM D1518-14 test method (ASTM D1518-14, 2014) covers only the measurement of the thermal resistance of the fabrics under steady-state conditions or dry heat flow (R_{ct} value), while ISO 11092:2014 (ISO 11092:2014) covers both the measurement of the thermal resistance and the water vapour resistance (R_{eT} value). ASTM D1518-14 (ASTM D1518-14, 2014) is limited to determinations of heat transfer

TABLE 9.3
The Overview of Other Significant ISO Standards in the Field of Thermal Measurements, Defining the Thermal Environment of Fabrics and Thermal Request Set upon Textiles and Clothing

Designation	Title
ISO 7243:1989	*Hot Environments – Estimation of the Heat Stress on Working Man, Based on the WBGT Index (Wet Bulb Globe Temperature)*
ISO 11399:1995	*Ergonomics of the Thermal Environment – Principles and Application of Relevant International Standards*
ISO 12894:2001	*Ergonomics of the Thermal Environment – Medical Supervision of Individuals Exposed to Extreme Hot or Cold Environments*
ISO 13731:2001	*Ergonomics of the Thermal Environment – Vocabulary and Symbols*
ISO 13732-1:2006	*Ergonomics of the Thermal Environment – Methods for the Assessment of Human Responses to Contact with Surfaces – Part 1: Hot Surfaces*
ISO/TS 13732-2:2001	*Ergonomics of the Thermal Environment – Methods for the Assessment of Human Responses to Contact with Surfaces – Part 2: Human Contact with Surfaces at a Moderate Temperature*
ISO 13732-3:2005	*Ergonomics of the Thermal Environment – Methods for the Assessment of Human Responses to Contact With Surfaces – Part 3: Cold Surfaces*
ISO/TS 14505-1:2007	*Ergonomics of the Thermal Environment – Evaluation of Thermal Environments in Vehicles – Part 1: Principles and Methods for Assessment of Thermal Stress*
ISO 14505-2:2006	*Ergonomics of the Thermal Environment – Evaluation of Thermal Environments in Vehicles – Part 2: Determination of Equivalent Temperature*
ISO 14505-3:2006	*Ergonomics of the Thermal Environment – Evaluation of Thermal Environments in Vehicles – Part 3: Evaluation of Thermal Comfort Using Human Subjects*
ISO/TS 14415:2005	*Ergonomics of the Thermal Environment – Application of International Standards to People with Special Requirements*
ISO 15265:2004	*Ergonomics of the Thermal Environment – Risk Assessment Strategy for the Prevention of Stress or Discomfort in Thermal Working Conditions*
ISO 15743:2008	*Ergonomics of the Thermal Environment – Cold Workplaces – Risk Assessment and Management*

from a warm, dry, constant-temperature, horizontal flatplate up through a layer of the test material to a cool atmosphere. In addition, although both ISO 11092 (ISO 11092:2014) and ASTM D1518-14 (ASTM D1518-14, 2014) can be used to calculate the resistance of the material with a horizontal airflow over the specimen (with conditioned air ducted to flow across and parallel to sample upper surface), ASTM D1518-14 (ASTM D1518-14, 2014) also provides the measurements made under still air conditions.

The ASTM D1518-14 (ASTM D1518-14, 2014) test method is limited to determinations on specimens of battings and layered batting/fabric assemblies having an intrinsic thermal resistance from 0.1 to 1.5 Km^2/W and thicknesses not in excess of 50 mm, contrary to ISO 11092:2014, which is restricted to a maximum thermal

TABLE 9.4
The Overview of ASTM Standards in the Field of Thermal Environment Measurements, Measurements of Thermal Characteristics of Fabrics and Clothing and Thermal Comfort of Human Subjects

Designation	Title
ASTM D1518-14	*Standard Test Method for Thermal Transmittance of Textile Materials*
ASTM F1868-17	*Standard Test Method for Thermal and Evaporative Resistance of Clothing Materials Using a Sweating Hot Plate*
ASTM D7024-04	*Standard Test Method for Steady State and Dynamic Thermal Performance of Textile Materials*
ASTM F1291-16	*Standard Test Method for Measuring the Thermal Insulation of Clothing Using a Heated Manikin*
ASTM F2370-16	*Standard Test Method for Measuring the Evaporative Resistance of Clothing Using a Sweating Manikin*
ASTM F237116	*Standard Test Method for Measuring the Heat Removal Rate of Personal Cooling Systems Using a Sweating Heated Manikin*

resistance and water vapour resistance which depends on the dimensions and construction of the apparatus used (e.g. $R_{ct} = 2$ m^{2}K/W and $R_{et} = 700$ m^{2}Pa/W respectively, for the minimum specifications of the equipment).

For the purposes of measuring both the thermal and the evaporative resistance of materials, another ASTM method should be used. This is the ASTM F1868-17 *Standard Test Method for Thermal and Evaporative Resistance of Clothing Materials Using a Sweating Hot Plate* (ASTM F1868-17, 2017).

When establishing the applicability of a certain personal protective clothing ensemble for a specific occupation, it is important to quantify the thermal resistance and evaporative resistance of textile fabrics, battings and other materials used for the production of those ensembles. The ASTM F1868-17 (ASTM F1868-17, 2017) method covers measurement of the thermal resistance and the evaporative resistance, under steady-state conditions, of fabrics, films, coatings, foams and leathers, including multilayer assemblies, for use in clothing systems. The range of this measurement technique for intrinsic thermal resistance is from 0.002 to 0.5 Km^2/W and for intrinsic evaporative resistance is from 0.01 to 1.0 kPa m^2/W (ASTM F1868-17, 2017).

There is a significant difference in air speed value proposed in official standards when measuring the thermal resistance of the fabric materials with the hotplate apparatus (ISO 11092:2014; ASTM D1518-14, 2014; ASTM F1868-17, 2017) and the thermal insulation of clothing with the help of the thermal manikins (ISO 15831:2004; ASTM F1291-16, 2016; ASTM F2370-16, 2016). The relatively laminar and horizontal air flow over the sweating guarded-hotplate proposed by ISO 11092:2014 is 1±0.05 m/s (3.6 km/h), measured 15 mm above the measuring unit (ISO 11092:2014). ASTM D1518-14 proposes still air conditions with air flow less than 0.1 m/s (0.36 km/h) (ASTM D1518-14, 2014). ASTM F1868-17 proposes air flow over the specimen of 1 m/s (3.6 km/h) (ASTM F1868-17, 2017). In thermal manikin measurements, the air speed is set to 0.4±0.1 m/s with the spatial variations not exceeding ±50% of the mean value within 0.5 m of the manikin surface (ISO 15831:2004). So the main question is how to compare the results of fabric thermal resistance to clothing thermal resistance if they have been measured under different climatic conditions. Havenith et al. proved the significant impact of body movement and wind on the clothing insulation (basic clothing thermal insulation corresponds to fabric thermal resistance) and there is significant difference in air speed values between standardised thermal resistance measurements for fabrics and clothing (Havenith et al., 1990). Airspeed affects the surface air layer insulation the most. The total clothing insulation, I_T (thermal insulation from the body surface to the environment including all clothing, enclosed air layers and boundary air layer), reduces between 15 to 26% with the wind increasing from no wind conditions to 0.7 m/s and between 34 to 40% with the wind increasing to 4 m/s when compared to still no wind conditions (air speed less than 0.1 m/s). The intrinsic or basic clothing insulation, I_{cl} (thermal insulation from the skin surface to the outer clothing surface including enclosed air layers), reduces 17% with the wind increasing from no wind conditions to 0.7 m/s and for 6 to 19% with the wind increasing to 4 m/s. The surface air layer insulation decreases for between to 40% with the wind increasing to 0.7 m/s and between 62 to 70% with the wind increasing to 4 m/s. They have also proven that total insulation decreases up to 32%, the basic clothing insulation up to 30% and the boundary (surface) air layer insulation decreases up to 35% when walking under the speed of 0.9 m/s (3.24 km/h). The combination of posture, movement and the wind affects total insulation up to 53%, the basic insulation up to 36% and surface air layer insulation up to 80%. On the top of it all, the clothing material provides less than 40% of the insulation and the air entrapped inside the clothing provides more than 60% to the value of the intrinsic clothing insulation (Havenith et al., 1990). It is presumably to expect the same decrease in the resistance value for fabrics under different air speed values.

The determination of the overall thermal transmission coefficient (thermal transmittance) due to conduction for dry specimens of textile fabrics, battings and other materials and the determination of the temperature regulating factor (TRF) is evaluated according to standard ASTM D7024-04 *Standard Test Method for Steady State and Dynamic Thermal Performance of Textile Materials* (ASTM D7024-04, 2004). This method is applicable for quality and cost control during manufacturing and can be used to establish criteria for creating thermal and comfort parameters for textiles used in the clothing industry. The fabric is placed between a hotplate and two coldplates and exposed to the controlled constant or varying flux. To measure the steady-state thermal resistance (R-value) of the fabric, the controlled flux is constant and the test proceeds until a steady state is reached. To measure the TRF, the flux is varied sinusoidally with time.

The test method ASTM F1291-16 *Standard Test Method for Measuring the Thermal Insulation of Clothing Using a Heated Manikin* covers the determination of the insulation value of different garments or clothing ensembles (ASTM F1291-16, 2016). It is applicable for the determination of the insulation properties of clothing to dry heat transfer in static conditions, dependent on the fabric layers, the coverage of the body or to be exact the amount of body surface area covered by clothing and the impact of the ease allowance added to specific clothing ensemble (looseness or tightness of fit). The ensembles' insulation values can be used for models that predict the different physiological responses of people in different environmental conditions. This test method is used to determine the insulation values of clothing ensembles, but it also describes dry heat transfer from a heated manikin to a relatively calm and cool environment. The testing is performed only for particular clothing ensembles in static conditions with thermal manikin standing. The consequences with respect to air movement, are not addressed in this test method (ASTM F1291-16, 2016).

The determination of the evaporative resistance of clothing ensembles is covered in accordance with standard ASTM F2370-16 *Standard Test Method for Measuring the Evaporative Resistance of Clothing Using a Sweating Manikin* (ASTM F2370-16, 2016). The measurement of the resistance to evaporative heat transfer of clothing ensembles is measured by a static standing heated sweating thermal manikin. The standard can be used in a model formation that predicts the different physiological responses of people under different environmental conditions. This test method covers determination of the evaporative resistance of clothing ensembles. It describes the measurement of the resistance to evaporative heat transfer from a heated sweating thermal manikin to a relatively calm environment. It also specifies the sweating thermal manikin configuration, testing protocol and testing conditions (ASTM F2370-16, 2016). The clothing evaporative resistance could be calculated by the isothermal heat loss method or mass loss method, but Wang et al. stated that the heat loss method overestimated the heat stress. The mass loss method was suggested to be used for calculating the clothing evaporative resistance, or a correction on the real evaporative heat loss should be made. The evaporative resistance based on the heat loss method produces greater values than those determined by the mass loss method because part of the evaporative heat energy was taken from the environment. In order to avoid the combined effects of dry heat transfer, condensation and the heat pipe on clothing evaporative resistance, thermal manikin measurements should be performed in the isothermal condition (Wang et al., 2011).

The evaluation and comparison of *PCS's-Personal Cooling Systems*, which are dressed in combination with classic clothing or other specified clothing ensembles, is performed according to ASTM F2371-16 *Standard Test Method for Measuring the Heat Removal Rate of Personal Cooling Systems Using a Sweating Heated Manikin* (ASTM F2371-16, 2016). The method covers the objective property evaluation for different personal cooling systems' constructions because of the resistance to dry heat transfer by means of conduction, convection and radiation. It requires the usage of a sweating heated thermal manikin. It is particularly crucial to take into account sweating capability, as a potentially large fraction of heat dissipation is associated with evaporative cooling. It describes the evaporative heat removal rate and the duration of cooling provided by the personal cooling system, in order to assess the effectiveness. The resistance to evaporative heat loss of PCS's is measured with static sweating thermal manikins constructed to maintain a certain constant uniform temperature over the body surface (ASTM F2371-16, 2016).

The manikin tests for measuring clothing thermal insulation by the dry heat loss are almost the same between the ISO and the ASTM (ISO 15831:2004; ASTM

TABLE 9.5
The Comparison of the Ambient Conditions for ISO and ASTM Thermal Manikin Standards

	Air Temperature, t_a (°C)	Manikin Surface Temperature, t_s (°C)	Air Velocity, v_a (m/s)	Relative Humidity, RH (%)	Duration of the Test (min)
ISO 15831:2004	At least 12 degrees lower than t_a	34±0.2	0.4±0.1	30–70	20
ASTM F1291-16	At least 12 degrees lower than t_a	35±0.2°C	0.4±0.1	30–70	30
ASTM F2370-16	35±0.5°C	35±0.5°C	0.4±0.1	40±5	30
ASTM F2371-16	35±0.5°C	35±0.5°C	0.4±0.1	40±5	120

F1291-16, 2016). The only significant difference is that there is no ISO standard on the evaporative resistance measurements as there is for ASTM (ASTM F2370-16, 2016). The ambient air relative humidity in ISO 15831:2004 (ISO 15831:2004) and ASTM F1291-16 (ASTM F1291-16, 2016) is set between 30 and 70%, preferably 50%. But in ASTM F2370-16 (ASTM F2370-16, 2016), the relative humidity is set to 40±5%. The same as in ASTM F2371-16 (144). The skin temperature is 1 degree lower for ISO 15831:2004 (ISO 15831:2004), 307.15±0.2 K (34±0.2°C), then for ASTM F1291-16 (ASTM F1291-16, 2016) and ASTM F2370-16, 308.15±0.5 K (35±0.2/0.5°C) (ASTM F2370-16, 2016). However, although both ISO 15831:2004 (ISO 15831:2004) and ASTM F1291-16 (ASTM F1291-16, 2016) demand the air temperature should be at least 12 degrees lower than the manikin's mean temperature, ASTM F2370-16 (ASTM F2370-16, 2016) testing is done with no difference between the air and the manikin temperature or to say in isothermal conditions 308.15±0.5 K (35±0.5°C). Since there is no temperature difference between the skin and the surrounding air, dry heat exchange will not occur. The air velocity is the same in ISO and all of the ASTM standards (0.4±0.1 m/s). Duration of the testing, after the system has reached a steady-state condition, is different between ASTM and ISO standards, see Table 9.5. The spatial variations within 0.5 m from the manikin's surface, the temporal variations during the test period and the sensor accuracy is also compared, see Table 9.6.

9.3 *ASHRAE HANDBOOK OF FUNDAMENTALS*

The ASHRAE organisation has issued its last version of the *Handbook of Fundamentals* in 2017. In Chapter 8 of the *ASHRAE Handbook*, thermal comfort is defined.

Thermal comfort is defined by ASHRAE 55-2017 and the *ASHRAE Handbook of Fundamentals* as that condition of mind that expresses satisfaction with the thermal environment (ANSI/ASHRAE Standard 55-2017; *ASHRAE Handbook of Fundamentals*, 2017). ASHRAE 55 and ISO 7730 are the only standards that define

TABLE 9.6
The Comparison of the Spatial and Temporal Variations and the Required Sensor Accuracy for ISO and ASTM Thermal Manikin Standards

	ISO 15831:2004	ASTM F1291-16	ASTM F2370-16	ASTM F2371-16
The spatial variations within 0.5 m from the manikin's surface				
t_a (°C)	±1.0	±1.0	±1.0	±1.0
v_a (m/s)	±50	±50	±50	±50
RH (%)	±10	±5	±5	±5
The temporal variations during the test period				
t_a (°C)	±0.5	±0.5	±0.5	±0.5
v_a (m/s)	±20	±20	±20	±20
RH (%)	±10	±5	±5	±5
The sensor accuracy				
t_a (°C)	±0.15	±0.15	±0.15	±0.15
v_a (m/s)	±0.05	±0.05	±0.05	±0.05
RH (%)	±5	±5	±5	±5

local thermal comfort in an indoor environment (ANSI/ASHRAE Standard 55-2017; ISO 7730:2005). Thermoregulatory control mechanisms are unconscious and conscious, and humans consciously deduce conclusions on thermal comfort or discomfort based on direct stimulus from the skin and core of the body. In general, thermal comfort occurs when body temperatures are held within narrow ranges and when skin moisture is low or when the situation doesn't require substantial physiological efforts of the organism for environmental adaptation. As a consequence of the adaptation, the human body is capable of initiating a number of conscious behavioural actions to reduce discomfort. Some of the possible behavioural actions for reducing discomfort are altering clothing variety, altering intensity and type of activity, changing posture or location, opening or closing a window, etc. Surprisingly, despite regional climate conditions, living conditions and cultures that differ widely throughout the world, the temperature that people choose for comfort, such as clothing, activity, humidity and air movement has been found to be very similar.

The *ASHRAE Handbook of Fundamentals* in its Chapter 8 summarises the fundamentals on human thermoregulation and comfort in terms useful to the engineer for operating systems and designs for the comfort and health of building occupants (*ASHRAE Handbook of Fundamentals* 2017).

REFERENCES

ANSI/ASHRAE Standard 55-2017 Thermal Environmental Conditions for Human Occupancy. 2017. Atlanta, GA: American Society of Heating, Refrigerating and Air-Conditioning Engineers Inc., ISSN 1041-2336.

Anttonen, H., J. Niskanen, H. Meinander, V. Bartels, K. Kuklane, R. E. Reinertsen, S. Varieras and K. Soltyñski. 2004. Thermal manikin measurements – Exact or not? *International Journal of Occupational Safety and Ergonomics (JOSE)* 10(3): 291–300.

ASHRAE Handbook of Fundamentals (SI ed.). 2017. Atlanta, GA: American Society of Heating, Refrigerating, and Air Conditioning Engineers Inc., ISBN 1931862702.

ASTM D 1518-14 Standard Test Method for Thermal Resistance of Batting Systems Using a Hot Plate. 2014. Book of Standards Volume: 07.01. Philadelphia, PA: ASTM International.

ASTM D1518-85 Standard Test Method for Thermal Transmittance of Textile Materials. 2003. ASTM International.

ASTM D7024-04 Standard Test Method for Steady State and Dynamic Thermal Performance of Textile Materials. 2004. ASTM International.

ASTM F 1291-16 Standard Test Method for Measuring the Thermal Insulation of Clothing Using a Heated Manikin. 2016. Book of Standards Volume: 11.03. Philadelphia, PA: ASTM International.

ASTM F 1868-17 Standard Test Method for Thermal and Evaporative Resistance of Clothing Materials Using a Sweating Hot Plate Test. 2017. Book of Standards Volume: 11.03. Philadelphia, PA: ASTM International.

ASTM F 2370-16 Measuring the Evaporative Resistance of Clothing Using a Sweating Manikin. 2016. Book of Standards Volume: 11.03. Philadelphia, PA: ASTM International.

ASTM F 2371-16 Standard Test Method for Measuring the Heat Removal Rate of Personal Cooling Systems Using a Sweating Heated Manikin. 2016. Book of Standards Volume: 11.03. Philadelphia, PA: ASTM International.

Bach, A. J. E., I. B. Stewart, A. E. Disher and J. T. Costello. 2015a. A comparison between conductive and infrared devices for measuring mean skin temperature at rest, during exercise in the heat, and recovery. *PLoS ONE* 10(2): e0117907.

Bach, A. J. E., I. B. Stewart, G. M. Minett and J. T. Costello. 2015b. Does the technique employed for skin temperature assessment alter outcomes? A systematic review. *Physiological Measurement* 36: 27–51.

Bethea, D. and K. Parsons. 2002. The development of a practical heat stress assessment methodology for use in UK industry. Research Report 008. UK: Health and Safety Executive, ISBN: 0717625338.

Bogdan, A. and M. Zwolińska. 2012. Future trends in the development of thermal manikins applied for the design of clothing thermal insulation. *Fibres & Textiles in Eastern Europe* 20(4–93): 89–95.

BS 7915:1998 Ergonomics of the Thermal Environment – Guide to Design and Evaluation of Working Practices for Cold Indoor Environments. 1998. BSI.

BS EN 14058:2004 Protective Clothing – Garment for Protection Against Cool Environments. 2004. BSI.

BS EN 342:2004 Protective Clothing – Ensembles and Garment for Protection Against Cold. 2004. BSI.

BS EN 511:2006 Protective Clothing – Protective Gloves Against Cold. 2006. BSI.

Diller, K. R. J. W. Valvano and J. A. Pearce. 2000. Bioheat transfer. *The CRC Handbook of Thermal Engineering* Boca Raton, USA: CRC Press, ISBN 0-8493-9581-X.

DIN 33403-5:1997 Klima am Arbeitsplatz und in der Arbeitsumgebung – Teil 5: Ergonomische Gestaltung von Kältearbeitsplätzen. 1997. DIN.

Goldman, R. F. 2006. Thermal manikins, their origins and role. In: *Thermal Manikins and Modelling, 6th International Thermal Manikin and Modelling Meeting*, ed. J. Fan. Hong Kong: The Hong Kong Polytechnic University, 3–18, ISBN: 962-367-534-8.

Griefahn, B. 1998. Estimated insulation of clothing worn in cool climates (0–15°C) compared to required insulation for thermal neutrality (IREQ). In: *Proceedings of the 1st International Symposium on Problems with Cold Work*, ed. I. Holmér and K. Kuklane. Stockholm, Sweden: National Institute for Working Life, 45–47, ISBN 91-70454-83-3.

Havenith, G. 2005. Thermal conditions measurement, Chapter 60. In: *Handbook of Human Factors and Ergonomics Methods*, ed. N. Stanton, A. Hedge, K. Brookhuis, E. Salas and H. Hendrics. Boca Raton, FL: CRC Press, 552–573. ISBN 0-415-28700-6.

Havenith, G. and H. O. Nilsson. 2004. Correction of clothing insulation for movement and wind effects, a meta-analysis. *European Journal of Applied Physiology* 92: 636–640.

Havenith, G., I. Holmér and K. Parsons. 2002. Personal factors in thermal comfort assessment: Clothing properties and metabolic heat production. *Energy and Buildings* 34: 581–591.

Havenith, G., I. Holmér, E. A. Den Hartog, K. C. Parsons. 1999. Clothing evaporative heat resistance proposal for improved representation in standards and models. *Annals of Occupational Hygiene* 43(5): 339–346.

Havenith, G., R. Heus and W. A. Lotens. 1990. Resultant clothing insulation: A function of body movement, posture, wind, clothing fit and ensemble thickness. *Ergonomics* 33(1): 67–84.

Holmér, I. 1998. Evaluation of thermal stress in cold regions – A strain assessment strategy. In: *Proceedings of the 1st International Symposium on Problems with Cold Work*, ed. I. Holmér and K. Kuklane. Stockholm: Sweden National Institute for Working Life, 31–38, ISBN 91-70454-83-3.

Holmér, I. 2006. Protective clothing in hot environments. *Industrial Health*, 44(3): 404–413.

Holmér, I. 2009. Evaluation of cold workplaces: An overview of standards for assessment of cold stress. *Industrial Health* 47: 228–234.

Holmér, I., H. Nilsson, G. Havenith and K. Parsons. 1999. Clothing convective heat exchange proposal for improved prediction in standards and models. *Annals of Occupational Hygiene* 43(5): 329–337.

ISO. ISO/TC 38 Textiles. https://www.iso.org/committee/48148.html (accessed on October 12, 2017).

ISO/TC 94 Personal safety – Personal protective equipment. https://www.iso.org/committee/50580.html (accessed on October 12, 2017).

ISO. ISO/TC 94/SC 3 Foot protection. https://www.iso.org/committee/50592.html (accessed on October12, 2017).

ISO. ISO/TC 94/SC 13 Protective clothing. https://www.iso.org/committee/50626.html (accessed on October 12, 2017).

ISO 10551:1995 Ergonomics of the Thermal Environment – Assessment of the Influence of the Thermal Environment Using Subjective Judgment Scales. 1995. ISO-International Organization for Standardization. (also BS EN ISO 10551:2001).

ISO 11079:2007 Ergonomics of the Thermal Environment – Determination and Interpretation of Cold Stress When Using Required Clothing Insulation (IREQ) and Local Cooling Effects. 2007. ISO-International Organization for Standardization.

ISO 11092:2014 Physiological Effects – Measurement of Thermal And Water-Vapour Resistance under Steady–State Conditions (Sweating Guarded-Hotplate Test). 2014. ISO-International Organization for Standardization.

ISO 14415:2005 Ergonomics of the Thermal Environment – Application of International Standards to People with Special Requirements. 2005. ISO-International Organization for Standardization.

ISO 15265:2004 Ergonomics of the Thermal Environment - Risk Assessment Strategy for the Prevention of Stress or Discomfort in Thermal Working Conditions. 2004. ISO-International Organization for Standardization.

ISO 15831:2004 Clothing – Physiological Effects – Measurement of Thermal Insulation by Means of a Thermal Manikin. 2004. ISO-International Organization for Standardization.

ISO 20344:2011 Personal Protective Equipment – Test Methods for Footwear. 2011. ISO-International Organization for Standardization.

ISO 28802:2012 Ergonomics of the Physical Environment – Assessment of Environments by Means of an Environmental Survey Involving Physical Measurements of the Environment and Subjective Responses of People. 2012. ISO-International Organization for Standardization.

ISO 7243:2017 Ergonomics of the Thermal Environment – Assessment of Heat Stress Using the WBGT (Wet Bulb Globe Temperature) Index. 2017. ISO-International Organization for Standardization. (BS EN 27243:2017).

ISO 7726:1998 Ergonomics of the Thermal Environment – Instruments for Measuring Physical Quantities. 1998. ISO-International Organization for Standardization.

ISO 7730:2005 Ergonomics of the Thermal Environment – Analytical Determination and Interpretation of Thermal Comfort Using Calculation of the PMV and PPD Indices and Local Thermal Comfort Criteria. 2005. ISO-International Organization for Standardization.

ISO 7933:2004 Ergonomics of the Thermal Environment – Analytical Determination and Interpretation of Heat Stress Using Calculation of Predicted Heat Strain. 2004. ISO-International Organization for Standardization.

ISO 811:2018 Textiles – Determination of Resistance to Water Penetration – Hydrostatic Pressure Test. 2018. ISO-International Organization for Standardization.

ISO 8996:2004 Ergonomics of the Thermal Environment – Determination of Metabolic Rate. 2004. ISO-International Organization for Standardization.

ISO 9237:1995 Textiles – Determination of the Permeability of Fabrics to Air. 1995. ISO-International Organization for Standardization.

ISO 9886:2004 Ergonomics – Evaluation of Thermal Strain by Physiological Measurements. 2004. ISO-International Organization for Standardization.

ISO 9920:2007 Ergonomics of the Thermal Environment – Estimation of Thermal Insulation and Water-Vapour Resistance of a Clothing Ensemble. 2007. ISO-International Organization for Standardization.

ISO 17699:2003 Footwear – Test Methods for Uppers and Lining – Water-Vapour Permeability and Absorption. 2003. ISO-International Organization for Standardization.

ISO 17702:2003 Footwear – Test Methods for Uppers – Water Resistance. 2003. ISO-International Organization for Standardization.

ISO 17705:2003 Footwear – Test Methods for Uppers, Lining and Insocks – Thermal Insulation. 2003. ISO-International Organization for Standardization.

ISO 22649:2016 Footwear – Test Methods for Insoles and Insocks – Water Absorption and Desorption. 2016. ISO-International Organization for Standardization.

ISO 20877:2011 Footwear – Test Methods for Whole Shoe – Thermal Insulation. 2004. ISO-International Organization for Standardization.

ISO 28803:2012 Ergonomics of the Physical Environment – Application of International Standards to People with Special Requirements 2012. ISO-International Organization for Standardization.

James, C. A., A. J. Richardson, P. W. Watt and N. S. Maxwell. 2014. Reliability and validity of skin temperature measurement by telemetry thermistors and a thermal camera during exercise in the heat. *Journal of Thermal Biology* 45: 141–149.

Kuklane, K., C. Gao and I. Holmér. 2007. Calculation of clothing insulation by serial and parallel methods: Effects on clothing choice by IREQ and thermal responses in the cold. *International Journal of Occupational Safety and Ergonomics (JOSE)* 13(2): 103–116.

Lymberis, A. and S. Olsson, 2003. Intelligent biomedical clothing for personal health and disease management: State of the art and future vision. *Telemedicine Journal and e-Health* 9(4): 379–386.

Malchaire, J., B. Kampmann, G. Havenith, P. Mehnert and H. J. Gebhardt. 2000a. Criteria for estimating acceptable exposure times in hot working environments: A review. *International Archives of Occupational and Environmental Health* 73: 215–220.

Malchaire, J., A. Piette, B. Kampmann, G. Havenith, P. Mehnert, I. Holmér, H. Gebhardt, B. Griefahn, G. Alfano and K.C. Parsons. 2000b. Development and validation of the Predictive Heat Strain (PHS) model. In: *Environmental Ergonomics IX*, ed. J. Werner and M. Hexamer. Aachen, Germany: Shaker Verlag, ISBN 3-8265-7648-9.

Malchaire, J., B. Kampmann, H. J. Gelshardt, P. Mehnert and G. Alfano. 1999. The Predicted Heat Strain Index: Modifications brought to the required sweat rate index. In: *The Book of Proceedings of the BIOMED 'HEAT' Research Project Conference: Evaluation and Control in Warm Working Conditions*, ed. J. Malchaire. Barcelona, Spain: BIOMED.

Malchaire, J., B. Kampmann, P. Mehnert, H. Gebhardt, A. Piette, G. Havenith. I. Holmér, K. C. Parsons, G. Alfano and B. Griefahn. 2002. Assessment of the risk of heat disorders encountered during work in hot conditions. *International Archives of Occupational and Environmental Health* 75: 153–162.

Mehnert, P., J. Malchaire, B. Kampmann, A. Piette, B. Griefahn and H. Gebhardt. 2000. Prediction of the average skin temperature in warm and hot environments. *European Journal of Applied Physiology* 82: 52–60.

Parsons, K. C. 1998. Working practices in the cold: measures for the alleviation of cold stress. In: *Proceedings of the 1st International Symposium on Problems with Cold Work*, ed. I. Holmér and K. Kuklane. Stockholm, Sweden: National Institute for Working Life, 48–56, ISBN 91-70454-83-3.

Parsons, K. C. 1999. International standards for the assessment of the risk of thermal strain on clothed workers in hot environments. *Annals of Occupational Hygiene* 43(5): 297–308.

Parsons, K. C. 2003. Thermal comfort for special populations, special environments and adaptive modelling, Chapter 9. In: *Human Thermal Environments: The Effects of Hot, Moderate, and Cold Environments on Human Health, Comfort and Performance*, 2nd edition. London, UK: Taylor & Francis, ISBN 0-415-23793-9.

Parsons, K. C. 2006. Heat stress standard ISO 7243 and its global application. *Industrial Health* 44(3): 368–379.

Parsons, K. C. 2013. Occupational health impacts of climate change: Current and future ISO standards for the assessment of heat stress. *Industrial Health* 51: 86–100.

Parsons, K. C., G. Havenith, I. Holmer, H. Nilsson, J. Malchaire. 1999. The effects of wind and human movement on the heat and vapour transfer properties of clothing. *Annals of Occupational Hygiene* 43(5): 347–352.

Risikko, T., H. Anttonen and E. Hiltunen. 1998. Performance criteria for cold protective clothing. In: *Proceedings of the 1st International Symposium on Problems with Cold Work*, ed. I. Holmér and K. Kuklane. Stockholm, Sweden: National Institute for Working Life, 75–76, ISBN 91-70454-83-3.

Taylor, L., S. L. Watkins, H. Marshall, B. J. Dascombe and J. Foster. 2016. The impact of different environmental conditions on cognitive function: A focused review. *Frontiers in Psychology* 6. doi:10.3389/fphys.2015.00372.

Wang, F., G. Chuansi, K. Kuklane and I. Holmér. 2011. Determination of clothing evaporative resistance on a sweating thermal manikin in an isothermal condition: Heat loss method or mass loss method? *Annals of Occupational Hygiene* 55 (7): 775–783.

Index

3D scanning, 162

absolute pressure, 190, 191
absorption, 52, 55, 63, 64, 208
acceptability scales, 98
acetylcholine, 22, 30
adenosine diphosphate (ADP), 16
adenosine triphosphate (ATP), 15, 16, 18
adsorption, 55, 64, 74, 195
aesthetic comfort, 45
affective component, 98
afferent pathway, 32
air temperature (t_a) 11, 13, 14, 15, 17, 19, 20, 22, 37, 46, 49, 51, 55, 59, 72, 73, 75, 101, 111, 122, 150, 161, 184, 186, 202, 203 (*see also* temperature of ambient air or ambient temperature)
air velocity (v_a), 13, 15, 37, 49, 55, 65, 70, 72, 95, 96, 101, 122, 151, 186, 191, 199, 202, 215
Alambeta apparatus, 87
ambient water vapour pressure, 22, 67, 69, 70, 184, 189, 199
American Engineering Standards Committee (AESC), 145
American Society for Testing and Materials (ASTM), 122, 136, 146
American Society of Heating and Air-Conditioning Engineers (ASHAE), 137, 138
American Society of Heating and Ventilating Engineers (ASHAVE), 138
American Society of Heating, Refrigerating and Air Conditioning Engineers (ASHRAE), 122, 137
American Society of Refrigerating Engineers (ASRE), 137, 138
anemometer, 192
apocrine sweat gland, 22
apparent evaporative cooling efficiency, 72
apparent evaporative heat loss of the wet manikin, 72
area of the human body, 11
attitude scaling techniques, 97
Austrian Standards Institute (ASI), 135
autonomic nervous system (ANS), 29

ballistic, puncture- or cut-resistant protective clothing, 83
basal metabolic rate (BMR), 17, 18
basic evaporative heat transfer resistance of the clothing, 69, 70

basic standards, 121, 127
behavioural component, 98
behavioural methods, 84, 86
Belding, H., 61, 93, 148, 149
Bell, P. A., 149
bellows effect, 96, 161
Berkeley thermal comfort model, 97
Bernard, C., 24
bimetallic thermometers, 188
biological (microbial) protective clothing, 83
body core temperature (*see also* the temperature of body core), 10, 11, 15, 23, 26, 27, 28, 30, 31, 200, 202
body ventilation systems (BVS), 162
Brager, G. S., 149, 150, 151
Breathability, 46, 49, 87, 88, 105
British Engineering Standards Association, 145
British Standards Institution (BSI), 3, 135, 146, 158, 183
Bureau voor Normalisatie (NBN), 135

calorimetry, 16, 48, 66, 147, 193, 194
Cannon, W. B., 24, 25
central nervous system (CNS), 24, 28, 36
Chauncy electrically heated manikin, 148, 149
chemical protective clothing, 83
chemical thermogenesis, 18, 23
chemical, biological, radiological and nuclear (CBRN) protective clothing, 111
chimney effect, 162
chlorofibres polyvinyl chloride fibres, 103
chromic materials (colour change), 106
Cleveland heated flat plate, 148
clo unit, 59
clothing surface area factor, 57, 60, 69, 70, 94
clothing basic or intrinsic insulation, 57, 94, 159, 160
clothing comfort, 4, 45, 46, 52, 53, 97, 98, 100
clothing efficiency factor, 57
clothing ensemble, 51, 61, 69, 71, 74, 75, 84, 85, 128, 153, 154, 159, 161, 176, 177, 178, 179, 192, 195, 197, 203, 205, 207, 208, 212, 213, 214
coating, 53, 55, 74, 88, 102, 103, 104, 105, 106, 109, 110, 11, 204, 212
coefficient of skin wettedness (skin wettedness), 14, 22, 49, 51, 63, 67, 69
cognitive component, 98
cold-induced injury, 147
cold protective clothing, 74, 83, 85, 106, 108, 109, 110, 153, 205, 208, 209,

221

cold stress, 15, 18, 102, 121, 163, 173, 174, 176, 177, 178, 179, 183, 184, 193, 195, 196, 202, 203
Committee on Conformity Assessment (CASCO), 140
Committee on Consumer Policy (COPOLCO), 140
Committee to support developing countries (DEVCO), 140
Composites Research Advisory Group (CRAG), 158
condensation (chilled mirror) dew-point hygrometer, 189
conduction, 7, 10, 13, 14, 15, 19, 21, 43, 44, 48, 51, 52, 53, 55, 57, 58, 62, 65, 73, 74, 87, 92, 95, 147, 161, 187, 200, 201, 202, 210, 213, 214
conductive materials, 106
co-normative research (CNR), 125
controlled wear trials, 86
convection, 7, 8, 13, 14, 15, 19, 21, 22, 23, 43, 44, 48, 51, 52, 53, 55, 57, 60, 62, 63, 64, 67, 92, 94, 95, 96, 102, 147, 159, 160, 161, 162, 199, 200, 202, 210, 214
convective heat exposures, 83
convective heat transfer coefficient (h_c), 11, 14, 19, 20, 57, 58, 60, 92, 94, 198
cooling method, 90
Coolmax®, 105
coulometric electrolytic hygrometers, 190
cup anemometers, 192
cup method, 86, 88, 89
cutaneous vasoconstriction of skin blood vessels, 18

dacron thread, 190
de Dear, R. J., 61, 94, 149, 151
deadweight-tester, 191
decrease in thermogenesis, 23
deflecting vane anemometers, 192
dermis, 22, 33, 34, 35, 36
Deutsches Institut für Normung (DIN), 136, 152
dew-point sensor, 189
dew-point temperature (tdp), 14, 49, 184, 189
diffusion, 10, 22, 37, 44, 52, 55, 63, 64, 66, 67, 68, 75, 87, 208
direct calorimetry, 16, 48, 66, 194
DISC scale, 50
Doppler velocimeters, 192
draft standard (prEN), 126
DuBois and Dubois surface area, 11, 49
Dudley, C. B., 136, 146
dynamic clothing insulation ($I_{Tdynamic}$), 198
dynamic pressure, 190
dynamic vapour permeability index ($i_{m,dynamic}$), 199
dynamic vapour resistance ($R_{Tdynamic}$), 44, 199

eccrine sweat gland, 21, 22
ectotherms, 23
effective latent heat of evaporation, 72, 73
effective thermal insulation of clothing ensemble (I_{cle}), 61
effective thermal insulation of clothing garment (I_{clu}), 61
efferent pathway, 32
efficiency of the muscle contraction, 19
Einstein law of energy conservation, 43
electrical heating garments (EHG), 112
electromechanical transducers, 190, 191
emissivity, 11, 23, 94, 96, 109, 148, 162, 187
endotherms, 7
Engineering Standards Committee, 135, 145
entropy, 9
environmental parameters, 13, 55, 101, 122
Environmental Research Division, 149
epidermis, 22, 33, 34, 36
epinephrine, 18, 22, 30, 36
ergonomics of the physical environment, 121, 123, 124, 152, 183, 193
ergonomics of the thermal environment, 121, 122, 140, 146, 150, 152, 153, 153, 174, 176, 177, 178, 179, 183, 185, 192, 193, 195, 196, 197, 202, 203, 210, 211
European agency for safety and health at work (EU–OSHA), 154
European Committee for Electrotechnical Standardization (CENELEC), 126, 129
European Committee for Standardization (CEN), 122, 126, 129, 154
European Telecommunications Standards Institute (ETSI), 126
evaluation scales, 98
evaporation, 7, 10, 12, 13, 14, 15, 19, 20, 21, 22, 48, 51, 52, 53, 54, 63, 64, 65, 72, 73, 74, 75, 91, 102, 199, 202, 209
evaporative heat transfer coefficient (h_e), 22, 69, 198
exergy, 8, 9, 10, 12
exergy analysis, 8, 12
exergy balance, 11
external environment, 24, 29, 46, 199

factors influencing the thermal insulation properties of textiles, 55
Fanger, P.O., 12, 13, 15, 46, 48, 50, 122, 150, 151, 155
feedback mechanisms, 23, 29, 30
feedforward mechanisms, 30
Fenn effect, 19
Fiala thermal comfort model, 97, 149
Fick's law of diffusion, 66
final draft, 125, 126, 141

Index

firefighter's clothing ensemble, 84
Fourier's law of thermal conduction, 58

Gauge pressure, 190
Gagge, A. P., 48
gas thermometer, 187
general cooling, 205
Gibbs-Dalton law of diffusion, 44
Givoni, B., 149
Goldbeater's skin, 190
Goldman, R. F., 149, 161
Goniometry, 90
Gore-Tex®, 88, 103, 105
Graphic Comfort Zone Method, 151
gravimetric hygrometer, 190
gross body mass loss, 68, 71, 202
guarded-hot plate, 87, 90, 91, 148
guarded hot plate method, 90

Hardy, J. D., 26, 48, 149
Harvard Fatigue Laboratory, 93, 147, 148
Havenith, G., 52, 70, 71
heat and flame protective clothing, 83, 103
heat balance equation of the human body (thermal balance), 12, 14, 15, 45, 148, 163, 203
heat capacity rate, 20, 21
heat flux, 12, 20, 58, 66 (*see also* thermal flux, heat flux density, heat transfer per unit area, heat flow per unit area or heat rate per unit area)
heat gain control mechanisms, 23
heat loss mechanisms, 23
heat of vaporization of water or enthalpy of evaporation, 67, 68, 69, 71
heat stress, 15, 50, 51, 71, 105, 121, 153, 156, 173, 174, 176, 177, 178, 196
heat transfer, 7, 8, 12, 19, 20
heat transmission pathways, 19
heating power supply, 59, 70
heating, ventilation and air conditioning systems (HVAC), 121
Herrington, L. P., 48, 147
Holmér, I., 161
homeostasis, 23, 24, 25, 26
homeotherms, 7, 24
hot cylinder method, 90
hot guarded semi-cylinder method, 90
hot disc method, 87, 90
Huizenga, C., 149
human thermoregulatory system, 7, 9, 13, 27, 95
humid operative temperature, 49, 51, 60
humidity ratio (W_a), 49, 184
Humphreys, M., 149, 151
hydrophilic solid membranes, 104
hygrometer, 189, 190

hygrometer that measure the colour changes, 190
hygrometer with electrical sensors, 190
hygrometer with mass sensitive piezoelectric sensors, 190
hygrometer with mechanical sensor, 190
hygrometer working on the basis of the evaporative cooling, 189
hypothalamus, 18, 23, 26, 28, 29, 30, 31, 32, 33, 36, 200, 203

increase in thermogenesis, 18
indirect calorimetry, 193
inspection testing standards, 127
Institute of Electrical and Electronic Engineers (IEEE), 127
internal environment, 7, 24, 25, 30
International Association for Testing Materials (IATM), 136, 137
International Electrotechnical Commission (IEC), 145
International Ergonomics Association (IEA), 152
International Federation of the National Standardizing Associations (ISA), 139, 145
International Organization for Standardization (ISO), 122, 138, 146
International Telecommunication Union (ITU), 146
International Telegraph Union, 146
interval scales, 98
intrinsic clothing insulation (I_{cl}), 60, 94, 96, 160, 198, 199, 210, 213

John B. Pierce Laboratory, 147, 149

Kata-thermometer, 192
Kawabata's KES-FB7 Thermolabo II system, 87
Kevlar®, 103
kilocalorie (kcal), 16
kilojoule (kJ), 16

latent or evaporative heat loss, 64
Lewis relation (L), 44, 67, 198
liquid-column manometers, 191
liquid-in-glass thermometers, 187
local cooling, 176, 177, 179, 202, 203, 205
low-pressure standard, 191
Lycra fibres, 105

magnetostrictive or electrostrictive materials, 107
management standards, 127
m-aramid fibres, 103
mass flow rate of blood per unit of time, 10, 11
mass flux or flow rate per unit area, 67

mass loss due to evaporation in the respiratory tract, 71
mass loss due to sweat loss, 71
mass loss due to the mass difference between carbon dioxide and oxygen, 72
mass variation due to variation of clothing or to sweat accumulation in the clothing, 72
mass variation of the body due to intake and excretion of water, 68, 72
mass variation of the body due to intake and excretions of solids, 68, 72
maximum evaporative potential, 68
maximum obtainable rate of heat flow, 21
maximum sweat rate 37
McCullough, E. A., 86, 149, 159, 160, 161
mean radiant temperature (t_r), 12, 13, 14, 49, 50, 55, 122, 185, 186, 187, 188, 193, 199, 202, 203
mechanical pressure gauges, 191
mechanical work, 12, 13, 14, 16, 19
metabolic rate of the body, 13, 14, 16, 17, 49, 199
metabolism, 8, 10, 12, 13, 15, 16, 17, 18, 19, 30, 37, 84, 97
metastatic thermometers, 188
microclimate heat pipe effect, 72, 74, 161
modelling methods, 84, 86
modes of vapour and moisture transport in porous materials, 63
moisture condensation, 52, 63, 64, 65, 71, 73, 74, 75, 161 (see also sweat)
moisture convection, 44, 63, 64
moisture permeation cylinder apparatus, 91
moisture vapour transmission rate (MVTR), 87, 89
MVTR cell method, 88

Nafion™, 112
Nanosphere®, 105
national standardizing bodies (NSB), 124
negative-feedback control system, 26
net heat production in human body, 13
Nevins, R. G., 122, 149
Newton's law of cooling, 44, 90
Nicol, J. F., 149, 151
Nishi, Y., 149
Nomex®, 103
nominal scales, 98
non-shivering thermogenesis, 18
norepinephrine, 18, 22, 30, 33, 36
novoloid or phenolic fibres, 103
nylon microfibers, 105

objective measuring methods, 84
observation time, 68
operative temperature (t_o), 49, 51, 55, 57, 60, 122, 150, 152, 186

ordinal scales, 98
outer shell, 7, 13, 110, 111
Outlast®, 106
oxygen uptake, 68, 193, 194

Pandolf, R. R., 149
parameters affecting the thermal comfort, 49
p-aramid fibres, 103
parasympathetic, 21, 29, 30
partial calorimetry, 48
partial pressure of the water vapour (absolute humidity of the air, p_a), 186
Patagonia BioMap®, 105
perceptual scales, 98, 195, 196
permeation coefficient (water vapour permeance coefficient of the skin), 67
Permetest method, 88
personal cooling systems (PCSs), 93, 162, 212, 214
personal heating garments (PHGs), 112, 162
personal parameters, 13
personal protective equipment (PPE), 122, 153, 154, 157, 176, 185, 206
phase change materials (PCM), 102, 106, 111
physical comfort, 45, 46
physiological comfort, 2, 45, 46, 102, 204
Pierce, J. B., 147, 149, 150
piezoelectric materials, 107
piloerection of the hair follicles, 18
Piston-gauge, 191
pitot-static tube, 192
pituitary gland, 18, 29, 30
Planck's law, 44
plane radiant temperature (t_{pr}), 185, 187, 188
poikilotherms, 7
Polartec® Power Stretch®, 105
polyacrylate fibres, 103
polyamide-imide fibres, 103
polybenzimidazole fibres, 103
polyester fibres, 56, 103, 105
polyetheretherketone fibres, 103
polyimide fibres, 103
polyphenylene sulphide fibres (PPS), 103
polytetrafluoroethylene membrane, 103, 104
polyurethane membrane, 104
predicted heat strain model (PHS), 199
predicted mean vote (PMV), 12, 50, 122, 150, 193
predicted percentage of dissatisfied (PPD), 12, 122, 151, 155, 193
preference scales, 98, 195
pre-normative research (PNR), 125
pressure-balance, 191
pressure-bulb thermometers, 188
process standards, 127
product standards, 127, 135

Index

propeller or revolving vane anemometers, 192
protective clothing, 56, 64, 74, 75, 76, 83, 84, 85, 92, 101, 103, 109, 110, 111, 127, 153, 154, 158, 163, 175, 185, 205, 206, 207
psychological comfort, 45, 46
psychrometer, 189
publicly available specification (PAS), 125
pyrometers or infrared radiation thermometers, 188

radiation, 7, 11, 13, 14, 15, 19, 22, 43, 44, 48, 51, 52, 53, 55, 57, 92, 109
radiative heat transfer coefficient (h_r), 23, 57, 60, 92, 198
rate at which regulatory sweat is generated, 68
rate of heat flow, 10, 21
rate of heat flow between the body core and the skin, 10
rate of heat loss by conduction, 13, 14
rate of heat loss by convection, 13, 14
rate of heat loss by evaporation (evaporative heat loss), 13, 14
rate of heat loss by radiation (radiative heat loss), 13, 14
rate of heat loss from skin, 14
rate of heat loss through respiration, 14
rate of heat storage in core compartment, 14
rate of heat storage in skin compartment, 14
rate of heat storage in the body, 13, 14
rate of the skin blood flow, 21
rating scales, 86, 98, 99
ratio scales, 98, 99
real evaporative cooling efficiency of the body, 72
real evaporative heat loss from the skin, 72
rectal temperature (t_{re}), 148, 199
reference temperature of the environment, 11
regulatory sweating, 67, 68
relative evaporative locus, 75
relative humidity (r_h), 49, 50, 55, 59, 65, 70, 100, 101, 150, 184, 189, 190, 202, 204, 215
required clothing insulation (IREQ), 176, 177, 179, 185, 202
required sweat rate index (SW_{req}), 199
resistance thermometer, 188
resistance to dry heat transfer, 91, 93, 214
resistance to evaporative heat transfer, 91, 93, 214
respiration, 7, 10, 14, 19, 48, 53, 92, 196
respiratory exchange ratio (R), 16, 17
respiratory quotient (RQ), 16
respiratory weight loss, 68
resting metabolism, 37, 43
resultant basic clothing insulation, 60
resultant or dynamic total thermal insulation of clothing ($I_{t,r}$), 60
resultant total water vapour resistance, 70
Rohles, F. H., 45, 62, 149

saturation density, 67
self-heating salt-phase transition hygrometer or a heated electrical hygrometer, 189
semi-carbon fibres, 103
semi-permeable membrane, 104
sensible heat loss, 13, 14, 57
sensorial comfort, 45, 46
set-point temperature, 26
shape memory alloy (SMA), 102, 107, 158
shape memory materials (SMM), 106, 108
shape memory polymers (SMPs), 107, 108
shivering, 8, 13, 17, 18, 20, 23, 26, 32, 33
skin temperature (t_{sk}), 14, 20, 33, 36, 47, 49, 50, 51, 57, 65, 74, 97, 100, 148, 161, 199, 200, 201
smart textiles, 101, 105, 106, 109
somatosensory system, 33
sorption, 63, 64, 74, 75
specific heat of the blood, 10, 11, 21
spectroscopic (radiation absorption) hygrometer, 190
standardization, 3, 4, 122, 126, 129, 138, 146, 154
static pressure, 190
steady-state conditions, 11, 12, 24, 69, 74, 91, 176, 185, 204, 210, 212
steady-state methods, 87
Stefan–Boltzmann law, 44
stimuli sensitive polymers (SSPs), 107
Stolwijk, J. A. J., 33, 149
subcommittee (SC), 124, 140, 141, 152, 153, 183
subjective judgement scales, 98, 185, 195, 196
subjective methods, 84, 86, 173
Suomen Standardisoimisliitto (SFS), 135
surface temperature (t_s), 19, 36, 44, 55, 70, 95, 122, 186, 187, 188, 189, 202
sweat evaporation, 20, 21, 22, 26, 32, 33, 95
sweat gland, 10, 22, 37, 50, 95, 100, 197, 199
sweat rate (SW_p), 10, 22, 37, 50, 95, 100, 197, 199
sweating guarded hot plate, 70, 72, 93, 95, 214
sweating thermal manikin, 20, 21, 29, 30
sympathetic nervous system, 124

technical committee (TC), 140
Technical Management Board (TMB), 4, 124, 125, 127
technical specification (TS), 103
Teflon®, 57
temperature at the clothing surface, 11
temperature of the skin, 213
temperature regulating factor (TRF), 90
tensiometry, 90
testing standards, 127, 129
thermal comfort, 9, 10, 13, 43, 45, 46, 48, 49, 50, 51, 53, 62, 86, 91, 93, 121, 145
thermal conductance, 58, 63, 64, 83, 87, 89, 90, 106, 107, 110, 201

thermal conductivity, 13, 56, 173, 203
thermal equilibrium, 60, 94, 96
thermal manikin, 192
thermal or hot-wire anemometers, 76, 83, 103
thermal protective clothing, 36, 46, 51, 56, 58, 59, 65, 87, 91, 109, 159, 210, 211, 212, 213
thermal resistance, 188, 200, 201
thermocouples, 17, 18
thermogenic effect of food, 187, 188, 192, 201
thermometer, 2, 45, 55, 56,
thermophysiological comfort, 35
thermoreceptive senses, 26, 31, 32, 33, 34, 35, 36
thermoreceptors, 4, 7, 12, 15, 22, 23, 24, 26, 29, 32, 75, 173, 192, 200, 216
thermoregulation, 97
thermoregulatory mathematical models, 105
Thermostat®, 32
thermostatic control centre, 18
thyrotropin-releasing hormone, 17, 18, 19
thyroxine, 98, 195
tolerance scales, 49, 184
total atmospheric pressure (p_t), 68
total evaporative heat loss from skin, 69
total evaporative resistance of the clothing ensemble and surface air layer, 57
total insulation of clothing and boundary air layer, 59, 94
total thermal insulation of clothing with surface air layer (I_T), 70, 198
total water vapour resistance (R_eT), 65, 74
total wet heat loss, 87
transient (or non-steady-state) methods, 50
TSENS scale, 147, 149

U.S. Army Medical Armored Research Laboratory, 147
U.S. Army Quartermaster Climatic Research Laboratory, 148, 149

U.S. Army Research Institute for Environmental Medicine (USARIEM), 97
UC Berkeley thermal comfort model, 139
United Nations Standards Coordinating Committee (UNSCC), 111

vacuum, 70, 198, 199
vapour permeability index (i_m), 198
vapour resistance of the boundary air layer (R_a), 23
vasodilation of skin blood vessels, 24
vasomotor action, 21
volumetric flow rate, 63, 66, 67

water vapour concentration difference, 55, 56, 59, 87, 88, 89, 91, 184, 208
water vapour permeability, 67, 89
water vapour permeance (W_d), 22, 67, 69
water vapour pressure at the skin surface, 88
water vapour transmission rate (WTR), 88
waterproof breathable fabrics, 16
watt (W), 100
wear trial experimental technique, 45, 49
wearing comfort, 49, 184
wet-bulb temperature (t_{wb}), 63, 74, 90, 95
wetting, 20, 21
whole body heat loss coefficient (heat capacity rate), 46, 52, 56, 63, 64, 72, 74, 75, 76, 90, 91, 95, 102, 105, 109, 204
wicking, 49, 202
wind chill index (WCI), 48, 147
Winslow, C. E. A., 124, 125, 126, 141, 152, 153, 154, 183
working group (WG), 137, 146
World War I (WW I), 135, 139, 146
World War II (WW II), 105

X-BIONIC's Symbionic Membrane®, 127

Standards Index

ANSI/ASHRAE Standard 55-1966, 150
ANSI/ASHRAE Standard 55-1992, 150, 151
ANSI/ASHRAE Standard 55-2004, 151
ANSI/ASHRAE Standard 55-2010, 151
ANSI/ASHRAE Standard 55-2017, 15, 46, 51, 93, 122, 138, 150, 152, 215, 216
ASHRAE Handbook Fundamentals, 14, 44, 46, 49, 51, 57, 58, 66, 67, 68, 71, 150, 179, 184, 187, 188, 189, 190, 191, 192, 193, 215, 216
ASTM D 1518-14, 210, 211, 212
ASTM D 1518-85, 122, 210
ASTM D 3410/D 3410M-95, 158
ASTM D 7024-04, 212, 213
ASTM E 96 / E 96M-16, 88, 89
ASTM F 1291-16, 93
ASTM F 1868-02, 89
ASTM F 1868-17, 89, 91
ASTM F 2298-03(2009)e1, 89
ASTM F 2370-16, 93
ASTM F 2371-16, 93, 162
ASTM F 2732-16, 93
ASTM F 372-73, 87
ASTM F2300-10, 162

BS 3424-12:1996, 90
BS 5750-2:1987, EN 29002-1987, 139
BS 7209:1990, 88
BS 7915:1998, 203
BS EN 12280-1:1998, 90
BS EN 12280-3:2002, 90
BS EN 14058:2004, 175, 176, 179, 185, 205
BS EN 340:2003, 154
BS EN 342:2004, 175, 176, 185, 205, 207
BS EN 344-1:1993, 122, 154, 157
BS EN 407:2004, 157, 206
BS EN 420 (2003) + A1 (2009), 158
BS EN 511:2006, 122, 154, 158, 175, 176, 179, 185, 205, 206, 209

DIN 53924:1997-03, 90
DIN EN 13537:2002-11, 154

INSTA 352, 153
INSTA 353, 153
INSTA 354, 153
ISO 1:1975, 139
ISO 1:2002, 139
ISO 1:2016, 139

ISO 10551:1995, 175, 177, 179, 185, 195
ISO 11079:2007, 175, 176, 177, 179, 185, 202, 203, 209
ISO 11092:2014, 175, 176, 185, 204, 210, 212
ISO 11399:1995, 177, 178, 211
ISO 12572:2016, 86
ISO 12894:2001, 175, 176, 178, 211
ISO 13506-1:2017, 163, 206
ISO 13506-2:2017, 163
ISO 13688:2013, 154
ISO 13732-1:2006, 175, 179, 211
ISO 13732-2:2001, 175, 179
ISO 13732-3:2005, 175, 179, 211
ISO 14126:1999, 158
ISO 1419:2018, 90
ISO 14415:2005, 122, 156, 193
ISO 15265:2004, 175, 176, 178, 211
ISO 15384:2018, 163, 206
ISO 15496:2018, 88, 89, 184
ISO 15743:2008, 175, 177, 179, 211
ISO 15831:2004, 175, 176, 185, 193, 208, 215, 216
ISO 17699:2003, 206, 207
ISO 17702:2003, 206, 207
ISO 17705:2003, 206, 207
ISO 20344:2004, 154
ISO 20344:2011, 157, 175, 176, 185, 206
ISO 20345:2011, 122, 154, 157
ISO 20346:2014, 122, 154, 157
ISO 20347:2012, 122, 154, 157
ISO 20877:2011, 207
ISO 22649:2016, 207
ISO 2528:2017, 91
ISO 28802:2012, 183
ISO 28803:2012, 193
ISO 31:1960, 139
ISO 5085-1:1989, 158
ISO 5085-2:1990, 158
ISO 7243:2017, 175, 176, 178, 193, 196, 200
ISO 7726:1998, 175, 176, 177, 178, 185
ISO 7730:1994, 151
ISO 7730:2005, 175, 176, 177, 179, 185, 192
ISO 7933:2004, 175, 176, 177, 178, 185, 193, 197, 200, 204
ISO 80000-1:2009, 139
ISO 811:2018, 184, 209
ISO 8996:2004, 175, 176, 177, 178, 185, 193
ISO 9000:2015, 139
ISO 9001:2015, 139
ISO 9002:1987, 139

ISO 9004:2009, 139
ISO 9004:2018, 139
ISO 9237:1995, 209
ISO 9886:2004, 175, 176, 177, 178, 185, 200

ISO 9920:2007, 175, 176, 178, 185, 195
ISO/FDIS 21420, 158
ISO/IEC Guide 2:2004, 3
ISO/R 1:1951, 139